AQA GCSE (9-1)
Chemistry
Grade 8/9 Booster Workbook

Dorothy Warren
Gemma Young

William Collins' dream of knowledge for all began with the publication of his first book in 1819. A self-educated mill worker, he not only enriched millions of lives, but also founded a flourishing publishing house. Today, staying true to this spirit, Collins books are packed with inspiration, innovation and practical expertise. They place you at the centre of a world of possibility and give you exactly what you need to explore it.

Collins. Freedom to teach

HarperCollins Publishers
The News Building
1 London Bridge Street
London SE1 9GF

**Browse the complete Collins catalogue at
www.collins.co.uk**

First edition 2016

10 9 8 7 6 5 4

© HarperCollins Publishers 2016

ISBN 978-0-00-819434-5

Collins® is a registered trademark of HarperCollins Publishers Limited

www.collins.co.uk

A catalogue record for this book is available from the British Library

Commissioned by Joanna Ramsay
Project managed by Sarah Thomas and Siobhan Brown
Copy edited by Tony Clappison
Proofread by Karen Roberts
Answer check by Gerard Delaney
Typeset by Jouve India Pvt Ltd.,
Artwork by Jouve India Pvt Ltd.
Cover design by We are Laura and Jouve
Cover images: Spyros/Shutterstock, Magnesium burning: Science Photo Library
Printed by Grafica veneta S.p.A

Contents

Introduction

This workbook will help you build your confidence in answering Chemistry questions for GCSE Chemistry and GCSE Combined Science.

It gives you practice in using key scientific words, writing longer answers, answering synoptic questions as well as applying knowledge and analysing information.

You will find all the different question types in the workbook so you can get plenty of practice in providing short and long answers.

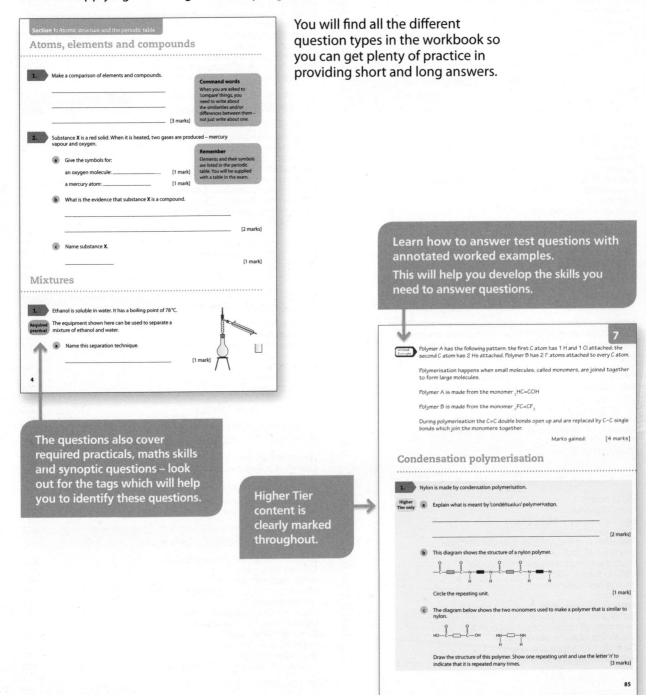

Learn how to answer test questions with annotated worked examples.

This will help you develop the skills you need to answer questions.

The questions also cover required practicals, maths skills and synoptic questions – look out for the tags which will help you to identify these questions.

Higher Tier content is clearly marked throughout.

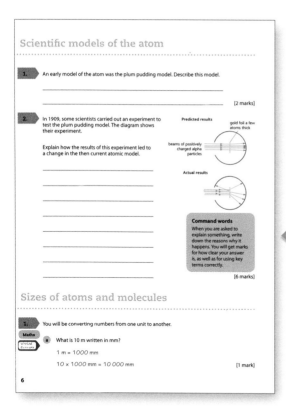

Scientific models of the atom

1. An early model of the atom was the plum pudding model. Describe this model.

_____

_____ [2 marks]

2. In 1909, some scientists carried out an experiment to test the plum pudding model. The diagram shows their experiment.

Explain how the results of this experiment led to a change in the then current atomic model.

Predicted results

gold foil a few atoms thick

beams of positively charged alpha particles

Actual results

Command words
When you are asked to explain something, write down the reasons why it happens. You will get marks for how clear your answer is, as well as for using key terms correctly.

_____

_____

_____

_____

_____

_____ [6 marks]

Sizes of atoms and molecules

1. You will be converting numbers from one unit to another.

Maths

a What is 10 m written in mm?

1 m = 1000 mm

10 × 1000 mm = 10 000 mm [1 mark]

6

There are lots of hints and tips to help you out. Look out for tips on how to decode command words, key tips for required practicals and maths skills, and common misconceptions.

The amount of support gradually decreases throughout the workbook. As you build your skills you should be able to complete more of the questions yourself.

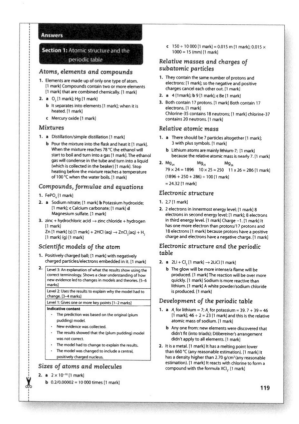

Answers

Section 1: Atomic structure and the periodic table

Atoms, elements and compounds

1. Elements are made up of only one type of atom. [1 mark] Compounds contain two or more elements [1 mark] that are combined chemically. [1 mark]
2. **a** O_2 [1 mark]; Hg [1 mark]
   b It separates into elements [1 mark]; when it is heated. [1 mark]
   c Mercury oxide [1 mark]

Mixtures

1. **a** Distillation/simple distillation [1 mark]
   b Pour the mixture into the flask and heat it [1 mark]. When the mixture reaches 78 °C the ethanol will start to boil and turn into a gas [1 mark]. The ethanol gas will condense in the tube and turn into a liquid (which is collected in the beaker) [1 mark]. Stop heating before the mixture reaches a temperature of 100 °C when the water boils. [1 mark]

Compounds, formulae and equations

1. $FePO_4$ [1 mark]
2. **a** Sodium nitrate; [1 mark] **b** Potassium hydroxide; [1 mark]; **c** Calcium carbonate; [1 mark] **d** Magnesium sulfate. [1 mark]
3. zinc + hydrochloric acid → zinc chloride + hydrogen [1 mark]
   Zn [1 mark] (s) [1 mark] + 2HCl (aq) → $ZnCl_2$ (aq) + H_2 [1 mark] (g) [1 mark]

Scientific models of the atom

1. Positively charged ball; [1 mark] with negatively charged particles/electrons embedded in it. [1 mark]
2.

Level 3: An explanation of what the results show using the correct terminology. Shows a clear understanding of how new evidence led to changes in models and theories. [5–6 marks]
Level 2: Uses the results to explain why the model had to change. [3–4 marks]
Level 1: Gives one or more key points [1–2 marks]
Indicative content
• The prediction was based on the original (plum pudding) model.
• New evidence was collected.
• The results showed that the (plum pudding) model was not correct.
• The model had to change to explain the results.
• The model was changed to include a central, positively charged nucleus.

Sizes of atoms and molecules

2. **a** 2×10^{-10} [1 mark]
   b 0.2/0.00002 = 10 000 times [1 mark]

c 150 ÷ 10 000 [1 mark] = 0.015 m [1 mark]; 0.015 × 1000 = 15 (mm) [1 mark]

Relative masses and charges of subatomic particles

1. They contain the same number of protons and electrons; [1 mark] so the negative and positive charges cancel each other out. [1 mark]
2. **a** 4 [1mark]; **b** 9 [1 mark]; **c** Be [1 mark]
3. Both contain 17 protons. [1 mark] Both contain 17 electrons. [1 mark] Chlorine-35 contains 18 neutrons; [1 mark] chlorine-37 contains 20 neutrons. [1 mark]

Relative atomic mass

1. **a** There should be 7 particles altogether [1 mark]; 3 with plus symbols. [1 mark]
   b Lithium atoms are mainly lithium-7; [1 mark] because the relative atomic mass is nearly 7. [1 mark]
2. Mg_{24} Mg_{25} Mg_{26}
   79 × 24 = 1896 10 × 25 = 250 11 × 26 = 286 [1 mark]
   (1896 + 250 + 286) ÷ 100 [1 mark]
   = 24.32 [1 mark]

Electronic structure

1. 2,7 [1 mark]
2. 2 electrons in innermost energy level; [1 mark] 8 electrons in second energy level; [1 mark] 8 electrons in third energy level. [1 mark] Charge –1. [1 mark] It has one more electron than protons/17 protons and 18 electrons [1 mark] because protons have a positive charge and electrons have a negative charge. [1 mark]

Electronic structure and the periodic table

2. **a** $2Li + Cl_2 \rightarrow 2LiCl$ [1 mark]
   b The glow will be more intense/a flame will be produced. [1 mark] The reaction will be over more quickly. [1 mark] Sodium is more reactive than lithium. [1 mark] A white powder/sodium chloride is produced. [1 mark]

Development of the periodic table

1. **a** A_r for lithium = 7; A_r for potassium = 39. 7 + 39 = 46 [1 mark]; 46 ÷ 2 = 23 [1 mark] and this is the relative atomic mass of sodium. [1 mark]
   b Any one from: new elements were discovered that didn't fit (into triads); Döbereiner's arrangement didn't apply to all elements. [1 mark]
2. It is a metal. [1 mark] It has a melting point lower than 660 °C (any reasonable estimation). [1 mark] It has a density higher than 2.70 g/cm³ (any reasonable estimation). [1 mark] It reacts with chlorine to form a compound with the formula XCl_2. [1 mark]

119

There are answers to all the questions at the back of the book. You can check your answers yourself or your teacher might tear them out and give them to you later to mark your work.

Atoms, elements and compounds

1. Make a comparison of elements and compounds.

_____

_____

_____ [3 marks]

Command words

When you are asked to 'compare' things, you need to write about the similarities and/or differences between them – not just write about one.

2. Substance **X** is a red solid. When it is heated, two gases are produced – mercury vapour and oxygen.

a Give the symbols for:

an oxygen molecule: _____ [1 mark]

a mercury atom: _____ [1 mark]

Remember

Elements and their symbols are listed in the periodic table. You will be supplied with a table in the exam.

b What is the evidence that substance **X** is a compound.

_____

_____ [2 marks]

c Name substance **X**.

_____ [1 mark]

Mixtures

1. Ethanol is soluble in water. It has a boiling point of 78 °C.

Required practical The equipment shown here can be used to separate a mixture of ethanol and water.

a Name this separation technique.

_____ [1 mark]

b Explain how to use it to separate a mixture of ethanol and water.

_____

_____

_____

_____ [4 marks]

Compounds, formulae and equations

· ·

1. Iron (III) phosphate contains four oxygen atoms and one phosphorus atom for every atom of iron.

What is its formula? Tick **one** box.

☐ Fe_3PO_4 ☐ $Fe(PO)_4$

☐ PO_4Fe ☐ $FePO_4$ [1 mark]

2. Name the following compounds:

a $NaNO_3$ _____ [1 mark]

b KOH_____ [1 mark]

c $CaCO_3$ _____ [1 mark]

d $MgSO_4$ _____ [1 mark]

> **Remember**
>
> Here are some tips for writing symbol equations:
>
> • Use the periodic table to look up the symbols of elements.
>
> • Elements that are gases at room temperature (other than the Group 0 elements) exist as pairs – for example, oxygen, O_2.
>
> • Symbol equations must be balanced. You can only change the numbers *in front* of the formulae.
>
> • You may be asked to provide state symbols (s, l, g, aq).

3. Metals react with acids to produce a salt and hydrogen.

Complete the word and symbol equations below. Write the state symbols in each equation. [5 marks]

zinc + hydrochloric acid → _____ + hydrogen   [1 mark]

_____ (___) + 2HCl (aq) → $ZnCl_2$(aq) + _____(___)   [4 marks]

Scientific models of the atom

1. An early model of the atom was the plum pudding model. Describe this model.

_____

_____ [2 marks]

2. In 1909, some scientists carried out an experiment to test the plum pudding model. The diagram shows their experiment.

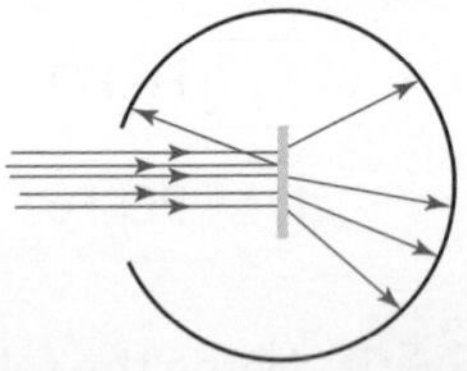

Explain how the results of this experiment led to a change in the then current atomic model.

_____

_____

_____

_____

_____

_____

_____

_____

Predicted results

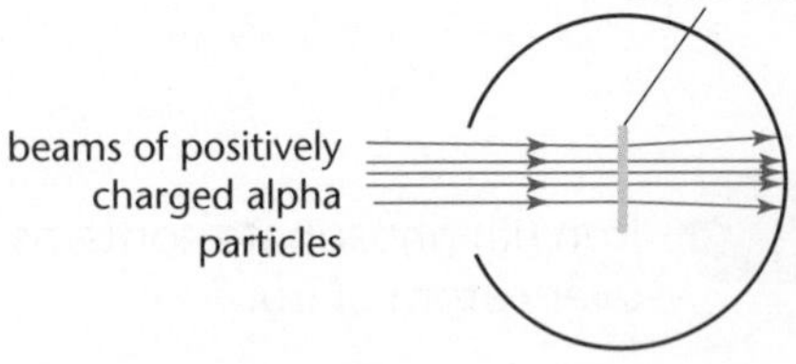

gold foil a few atoms thick

beams of positively charged alpha particles

Actual results

Command words

When you are asked to explain something, write down the reasons why it happens. You will get marks for how clear your answer is, as well as for using key terms correctly.

[6 marks]

Sizes of atoms and molecules

1. You will be converting numbers from one unit to another.

Maths

Worked Example

a What is 10 m written in mm?

1 m = 1000 mm

10 × 1000 mm = 10 000 mm

[1 mark]

6

b Write 5.2 mm in nm.

5.2 mm × 1000 × 1000 = 5 200 000 nm

[1 mark]

Marks gained: _____ [2 marks]

Maths

Metres (m) are used to measure length. **Kilo**metres, **milli**metres, **micro**metres and **nano**metres are other units of length that contain a prefix. A prefix indicates the size of the unit:

1 km = 1000 m

1 m = 1000 mm

1 mm = 1000 μm

1 μm = 1000 nm

You need to be able to convert numbers between units with prefixes.

2. Nanometres (nm) are a unit of measurement. 1 nm = 1×10^{-9} m.

a The diagram shows the diameter of a helium atom.

Write 0.2 nm in metres. Give your answer in standard form.

_____ [2 marks]

b The diagram below shows the diameter of the nucleus of the helium atom.

Use the information in the two diagrams to calculate how many times larger the diameter of the atom is than the diameter of the nucleus.

_____ [1 mark]

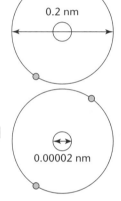

0.2 nm

0.00002 nm

c A teacher uses a circular sports stadium as **a scale model** of the helium atom.

The stadium has a diameter of 150 m. The nucleus is modelled by a sphere in the centre of the stadium. Calculate the diameter that the sphere needs to be in mm.

Maths

Because atoms are very small, standard form is used to show their size. You need to be able to use numbers in standard form.

Diameter of sphere = _____ mm [3 marks]

Relative masses and charges of subatomic particles

1. Atoms have no overall charge. Explain why.

_____

_____ [2 marks]

2. Study this diagram of an atom.

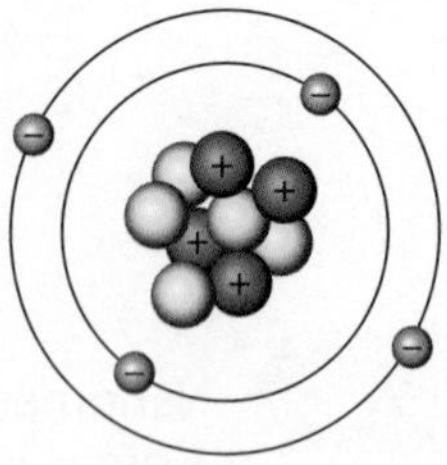

a What is its atomic number? _____ [1 mark]

b What is its mass number? _____ [1 mark]

c What is its symbol? _____ [1 mark]

3. These are two isotopes of chlorine.

$$^{35}_{17}Cl \qquad ^{37}_{17}Cl$$

Chlorine-35   Chlorine-37

Compare the numbers of subatomic particles in the atoms of each isotope.

_____

_____

_____

_____ [4 marks]

Relative atomic mass

1. There are two common isotopes of lithium. The diagram shows the nucleus of a lithium-6 atom.

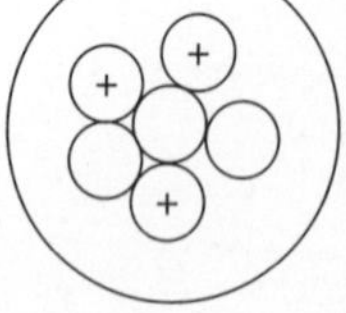

a In the space below, draw a diagram to show the nucleus of lithium-7.

[2 marks]

b The relative atomic mass of lithium is 6.941. Explain what this tells you about the relative abundances of each isotope.

_____

_____ [2 marks]

Common misconception

Not all the atoms of one element have the same mass number. The relative atomic mass is an average value that takes account of the abundance of the isotopes of the element.

2. The relative abundances of magnesium isotopes are 79% magnesium-24, 10% magnesium-25 and 11% magnesium-26

Maths

Calculate the relative atomic mass of magnesium. Give your answer to one decimal place.

Relative atomic mass of magnesium = _____ [3 marks]

Electronic structure

1. What is the electronic structure of a fluorine atom?

Tick **one** box.

☐ 7

☐ 2, 7

☐ 2, 8, 7

☐ 2, 8, 8, 7 [1 mark]

> **Analysing questions**
> To work out an electronic structure you need to know the atomic number of the element. This can be found on the periodic table.

2. Group 7 elements react with Group 1 elements. During the reaction, an electron is transferred from the Group 1 atoms to the Group 7 atoms.

In the space below draw a diagram to show the electronic structure of a chloride **ion**. Give the charge and explain why it has this charge.

_____

_____

_____

_____ [6 marks]

Electronic structure and the periodic table

1. The diagram shows the periodic table. Some elements are indicated by letters (these are not symbols).

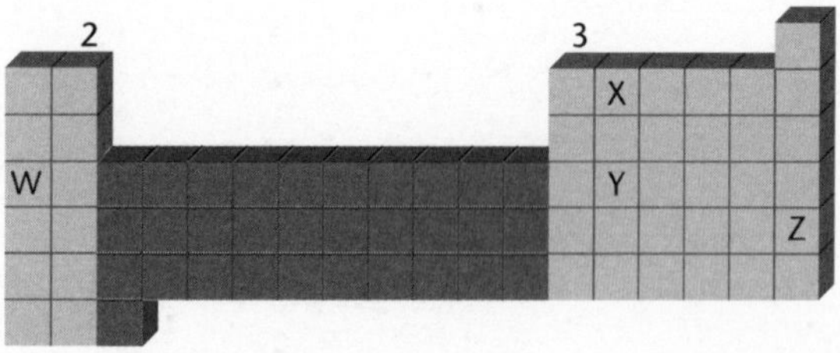

Worked Example Identify which **two** elements have similar chemical properties. Describe why this is so.

[3 marks]

X and **Y** [1 mark] *because they are in the same group/column. [1 mark] So they have the same number of electrons in their outer shell/both have 4 electrons in their outer shell. [1 mark]*

Marks gained: [3 marks]

2. A piece of lithium is put in a gas jar full of chlorine. The lithium glows and a white powder is produced. The reaction takes about 30 seconds to complete.

Synoptic **a** Write the symbol equation for this reaction.

_____ [2 marks]

b Predict what will happen when sodium is reacted with chlorine. Give reasons for your answers.

_____

_____

_____

_____ [4 marks]

Development of the periodic table

1. Johann Döbereiner worked on arranging the elements and by 1829 he had placed elements with similar chemical properties into groups of three – called triads. He noticed that the atomic weight of the middle element was approximately the mean of the other two.

Two examples are shown here.

| Calcium: 40 |
| Strontium: 88 |
| Barium: 137 |

| Chlorine: 35.5 |
| Bromine: 80 |
| Iodine: 127 |

Maths **a** Döbereiner suggested that lithium, sodium and potassium also formed a triad. Use the relative atomic masses given in the periodic table to explain why.

_____

_____

_____ [3 marks]

b Suggest why Döbereiner's ideas were replaced by those of Mendeleev.

_____

_____ [1 mark]

2. Building on Döbereiner's work, in 1869 Dmitri Mendeleev produced an early version of the periodic table. One feature of Mendeleev's work involved leaving gaps where there were breaks in the pattern he proposed.

An example of this can be seen in Group 3.

Row	Element	Metal or non-metal	Melting point in °C	Density in g/cm³	Formula of the chloride
1	Boron	Metal	2076	2.34	BCl_3
2	Aluminium	Metal	660	2.70	$AlCl_3$
3	**X**				

Predict the properties of the missing element, **X**, in Group 3.

_____

_____

_____

_____ [4 marks]

Analysing questions

You know that there are patterns in element properties as you go down a group – for example the melting points may increase going down a group.

Look at the patterns in the two named elements in the table and use the data to predict the properties of the missing element.

Comparing metals and non-metals

1. This diagram shows the electronic structure of an element.

Synoptic

   a   Name the element. _____   [1 mark]

   b   Explain why the element is a non-metal based on its electronic structure.

   _____

   _____   [2 marks]

2. The elements in the table are either metals or non-metals that do **not** have typical physical properties.

Element	Does it conduct electricity?	Melting point in °C	Density in g/cm³	Appearance
A (metal)	Yes	180.5	0.53	Shiny
B (non-metal)	Yes	5530	2.3	Dull
C (metal)	Yes	−38.8	13.6	Shiny

For each element, which property is not typical?

_____

_____   [3 marks]

Elements in Group 0

1. The table shows data about the Group 0 elements.

Maths

Element	Helium	Neon	Argon	Krypton	Xenon
Atomic number	2	10	18	36	54
Boiling point (°C)	−268.9	−246.1		−153.22	−108.1

a Plot the data on the axes below. Then draw a line of best fit.

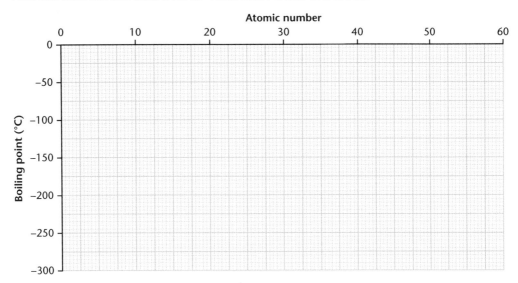

Atomic number

[3 marks]

b Estimate the boiling point of argon. _____

[1 mark]

2. Use electronic structure to explain why the elements in Group 0 of the periodic table are unreactive.

Maths
You can use a line of best fit to estimate values for missing data.

_____

_____ [3 marks]

Elements in Group 1

1. A teacher shows a class the Group 1 metals.

Required practical They cut a piece of lithium from a larger block. The shiny surface slowly goes dull because of an oxidation reaction.

Complete the **balanced** symbol equation to show this reaction.

_____ Li + _____ → _____ [3 marks]

2. Use the information in the table to answer the questions that follow.

Name of element	Atomic number	Melting point in °C	Density in g/cm³
Lithium	3	181	0.53
Sodium	11	98	0.97
Potassium	19	64	0.86
Rubidium	37	39	1.48

Maths **a** The diagram shows a lump of lithium and some data about it.

Calculate its mass.

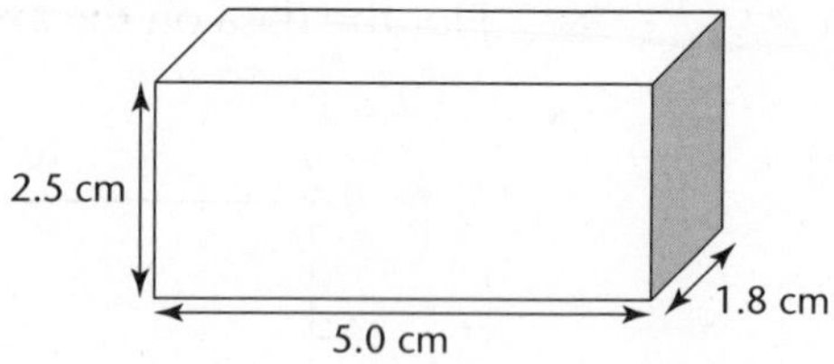

_____

2.5 cm

5.0 cm

1.8 cm

[2 marks]

b Describe the trends in physical properties of the Group 1 metals as atomic number increases.

_____

_____

_____ [3 marks]

Elements in Group 7

1.

Worked Example

Higher Tier only

Chlorine reacts with potassium bromide. Write the ionic equation for this reaction.

[2 marks]

The full equation is $Cl_2(g) + 2KBr(aq) \rightarrow 2KCl(aq) + Br_2(aq)$

So the ionic equation is $Cl_2(g) + 2Br^-$ (aq) [1 mark] $\rightarrow 2Cl^-(aq) + Br_2(aq)$ [1 mark]

Marks gained: [2 marks]

Analysing questions

1. Write out the full balanced symbol equation:

$Cl_2(g) + 2KBr(aq) \rightarrow 2KCl(aq) + Br_2(aq)$

2. Add charges onto the ions.

$Cl_2(g) + 2K^+2Br^-(aq) \rightarrow 2K^+2Cl^-(aq) + Br_2(aq)$

3. Cross out any ions that do not change.

$Cl_2(g) + \cancel{2K^+} 2Br^-(aq) \rightarrow \cancel{2K^+} 2Cl^-(aq) + Br_2(aq)$

4. Write out the completed ionic equation.

$Cl_2(g) + 2Br^-(aq) \rightarrow 2Cl^-(aq) + Br_2(aq)$

2.

Required practical

a A student bubbled chlorine gas through some colourless sodium iodide solution. The solution turned dark red. Explain why this happened.

_____

_____ [2 marks]

Higher Tier only

b Give an ionic equation for the reaction.

_____ [2 marks]

3. The diagrams show the electronic structures of the first three Group 7 elements. Use the diagrams to explain why the reactivity of the halogens decreases as you go down the group.

Fluorine 2, 7

Chlorine 2, 8, 7

Bromine 2, 8, 8, 7

_____

_____

_____ [6 marks]

Properties of the transition metals

1. The transition metal copper is commonly used to make pipes to carry water.

a Suggest **two** physical properties of copper that make it suitable for this function.

_____

_____ [2 marks]

> **Remember**
> Physical properties are those like hardness, strength, size and shape. They can be measured without carrying out a chemical reaction. Chemical properties relate to how substances react.

b Suggest **one** chemical property that makes it suitable for this function.

_____ [1 mark]

2. A compound has the formula $Co_2(SO_4)_3$

a Name the transition metal in the compound. _____ [1 mark]

b Determine the charge on the transition metal ion. _____ [1 mark]

3. The elements in Group 1 are also metals. They have very different properties from transition metals.

Compare the physical and chemical properties of Group 1 and transition metals.

_____

_____

_____ [6 marks]

The three states of matter

1.

Required practical

A student wanted to find out the melting point of a compound called salol. This is the method they used.

First put two spatulas of salol into a boiling tube and add a thermometer. Then stand the boiling tube in a hot water-bath.

a Describe how they would measure the melting point of salol.

_____

_____ [2 marks]

b They could not measure the melting point of magnesium oxide in school because its melting point is much higher than that of salol. Suggest why.

_____

_____ [2 marks]

2.

Higher Tier only

The three states of matter can be represented by particle diagrams.

Evaluate this model.

solid   liquid   gas

_____

_____

_____

Command words

When you are asked to 'evaluate' something you should use the information supplied as well as your knowledge and understanding to consider evidence for and against.

_____

So, in this question you should describe the arguments for and against how well this model represents real particles in the three states of matter.

_____

_____

_____

_____

[6 marks]

Ionic bonding and ionic compounds

1. Copper sulfate ($CuSO_4$) is an ionic compound.

a What ions does it contain? Tick **one** box.

☐ Cu^{2+}, S^{6+} and $4O^{2-}$ ☐ Cu^{2+}, S^{6-} and $4O^{2-}$

☐ Cu^{2+} and SO_4^{2-} ☐ Cu^{2+} and SO_4 [1 mark]

b Describe how you can tell that copper sulfate is an ionic compound from its name.

_____ [1 mark]

2. The ball and stick diagrams show the structures of two ionic compounds.

A  **B**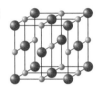

a Which diagram represents sodium oxide?

_____ [1 mark]

 b Explain your choice. [3 marks]

The formula of sodium oxide is Na_2O [1 mark] – you can work this out by looking on the periodic table. Sodium is in Group 1 so its ions have a +1 charge. Oxygen is in Group 6 so oxide ions have a –2 charge. In diagram A there are 14 of the large spheres and 8 of the smaller ones. [1 mark] This is a ratio of approximately 2 : 1, so it must contain 2 sodium ions for every 1 oxide ion. [1 mark]

Marks gained: [3 marks]

Remember

You can count the numbers of each type of ion to give you an estimate of the ratio. It won't be an exact number because these are just part of much bigger structures. The ions on the sides and corners are bonded to other ions not included in the image.

Dot and cross diagrams for ionic compounds

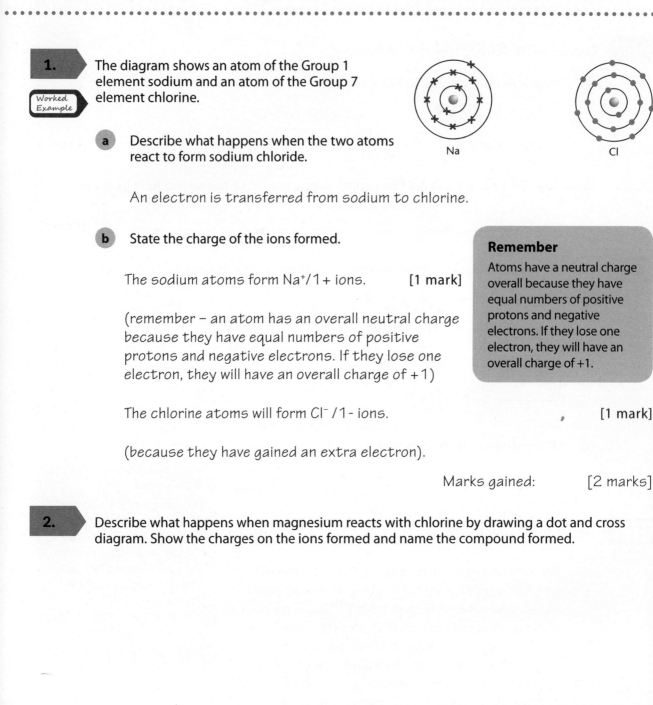

1. *Worked Example*

The diagram shows an atom of the Group 1 element sodium and an atom of the Group 7 element chlorine.

Na

Cl

a Describe what happens when the two atoms react to form sodium chloride.

An electron is transferred from sodium to chlorine.

b State the charge of the ions formed.

The sodium atoms form Na^+/1+ ions.   [1 mark]

(remember – an atom has an overall neutral charge because they have equal numbers of positive protons and negative electrons. If they lose one electron, they will have an overall charge of +1)

> **Remember**
> Atoms have a neutral charge overall because they have equal numbers of positive protons and negative electrons. If they lose one electron, they will have an overall charge of +1.

The chlorine atoms will form Cl^-/1- ions.   [1 mark]

(because they have gained an extra electron).

Marks gained: [2 marks]

2. Describe what happens when magnesium reacts with chlorine by drawing a dot and cross diagram. Show the charges on the ions formed and name the compound formed.

_____

_____

_____

_____   [6 marks]

Properties of ionic compounds

1. Look at the information in the table.

Ionic compound	Melting point in °C	Does it conduct electricity when solid?	Does it conduct electricity when molten?
Sodium chloride	804	No	Yes
Calcium oxide	2572	No	Yes

a Explain the difference in the conductivity of ionic compounds when solid and when molten.

_____

_____ [2 marks]

b Explain why ionic compounds have high melting points. Suggest a reason for the difference in melting points between the two compounds in the table.

Literacy

There are two parts to this question so answer one at a time. Aim to include six separate points for the 6 marks. Use scientific words in your answer – such as 'ions', 'forces of attraction', 'energy'.

_____

_____

_____

_____

_____

_____ [6 marks]

Covalent bonding in small molecules

1. Which substances are made up of small molecules?
   Tick **two** boxes.

   ☐ Ca   ☐ CO_2 ☐ Cl_2 ☐ $CaCl_2$ [2 marks]

Analysing questions
You need to be able to identify small molecules from their formulae. Small molecules are covalent (made up of non-metal atoms).

2. The diagram shows one molecule of a covalent substance.

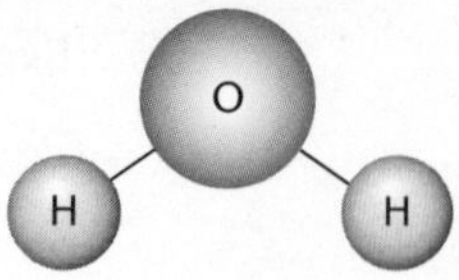

   a Name the substance. _____ [1 mark]

   b Describe how the bonds between the atoms are formed.

   _____ [1 mark]

3. Substance **Z** is made up of small molecules. The diagram represents one of these molecules.

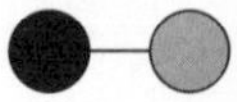

   a What is the force called that holds the atoms together?

   _____ [1 mark]

   Synoptic b The diagram below shows a substance in the liquid state.

   In the empty box, draw the molecules after the substance has been heated beyond its boiling point. [2 marks]

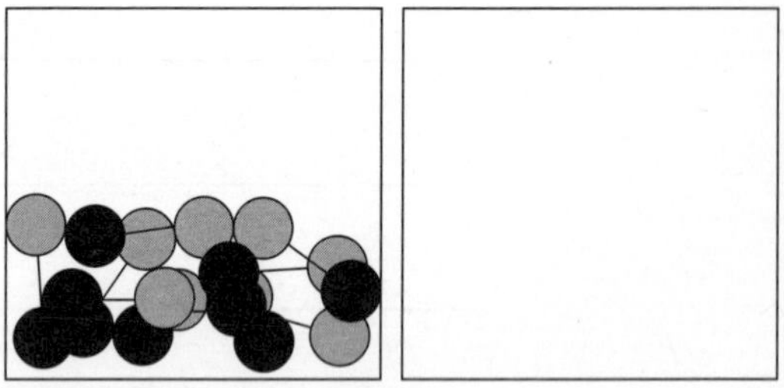

   c Predict how the boiling point of the substance in part (b) compares to that of sodium chloride.

   Give a reason for your answer.

   _____

   _____

   _____

   _____ [3 marks]

Dot and cross diagrams for covalent compounds

1. A dot and cross diagram is not a true representation of the structure of a small molecule. Give one reason why.

_____ [1 mark]

2. In the space below, draw a dot and cross diagram to show the covalent bonding in a molecule of ammonia (NH_3). Show the outer electron shells only.

> **Remember**
> In some molecules, more than one pair of electrons are shared.

[4 marks]

Properties of small molecule compounds

1. The alcohols are a family of molecules. Look at the data in the table.

Substance	Formula	Boiling point in °C
Methanol	CH_3OH	65
Ethanol	C_2H_5OH	79
Propanol	C_3H_7OH	97

> **Common misconception**
> It is important to remember that there are bonds between the atoms in a molecule and also _between_ the molecules.

Name the force that exists between the

a Atoms in an ethanol molecule: _____ [1 mark]

b Molecules of ethanol: _____ [1 mark]

2. The alcohols in the table above are all liquids at room temperature (25 °C).

Synoptic Explain, in terms of their bonding, why.

_____

_____ [2 marks]

3. Describe the trend in boiling point shown in the table above. Give an explanation for this trend.

_____

_____

_____

_____   [4 marks]

Polymers

1. The diagram shows one repeating unit of the polymer polystyrene. Complete it to show the structure of the polymer molecule.   [2 marks]

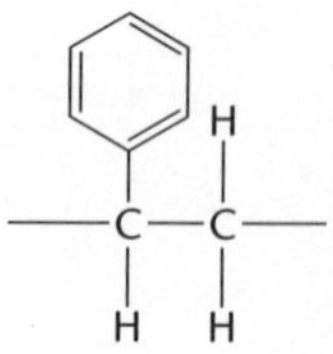

2. Poly(ethene) is a polymer of ethene.

Explain what this means.

_____   [1 mark]

Giant covalent structures

1. Simple molecules and giant covalent structures are both covalently bonded. Some examples are listed below.

water   **diamond**   **ammonia**   **silicon dioxide**

a Complete the table below by writing the names of these substances in the correct columns.

Simple molecules	Giant covalent molecules

[4 marks]

b Identify an example from your table that is

- an element: _____ [1 mark]

- a gas at room temperature: _____
  [1 mark]

- made up of carbon atoms only: _____
  [1 mark]

> **Synoptic** In the exam, some of the marks will be for connecting your knowledge from different areas of chemistry.

2. Both silicon dioxide (SiO_2) and carbon dioxide (CO_2) are covalent compounds. Compare their bonding and their structures.

_____

_____

_____

_____

_____ [4 marks]

Properties of giant covalent structures

1. Pure quartz is silicon dioxide.

Synoptic **a** Define 'pure'.

_____ [1 mark]

> **Command words**
> 'Define' means to state the meaning of something. So in this question you need to say what 'pure' means.

b Pure quartz and diamond are both electrical insulators. Identify **two** other physical properties that they share.

_____

_____ [2 marks]

c Explain why they share these properties

_____

_____ [2 marks]

2. The diagram shows the structures of diamond and graphite.

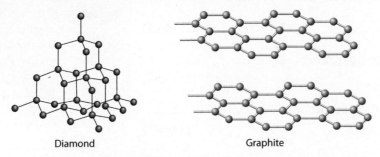

Diamond   Graphite

Compare the physical properties of diamond and graphite. Explain these properties in terms of their structures.

_____

_____

_____

_____

_____

_____

_____

_____ [6 marks]

Graphene and fullerenes

1. The diagram shows two types of carbon structures.

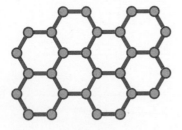

Graphene

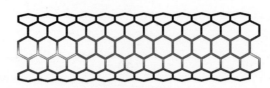

Carbon nanotube

a Which structure is a fullerene? Give a reason for your choice.

_____ [2 marks]

Synoptic **b** Carbon nanotubes are useful for making catalysts. Describe the function of a catalyst and explain why its properties make it suitable for this use.

_____

_____

_____

_____ [4 marks]

c Graphene is a single layer of graphite just one atom thick. It has delocalised electrons. One possible use for graphene is making touchscreens for electronic devices. Suggest how its properties make it suitable for this function.

_____

_____

_____

_____

_____ [6 marks]

Nanoparticles

1. The table shows data about different types of particles. Use the data to answer the questions that follow.

Maths

Particle	Diameter
Soot	1000 nm
Titanium dioxide	10 nm
Pollen grain	10 000 nm

Maths

If a number is about 10 times bigger than another, it is one order of magnitude bigger.

a How many orders of magnitude are there between soot particles and titanium dioxide particles?

_____ [1 mark]

b Which type of particle is classed as a nanoparticle? Give a reason for your answer.

_____

_____ [2 marks]

2. Gold can be used as a catalyst. The cube of gold shown here has 2 cm sides.

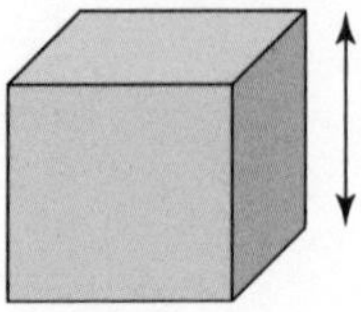

Maths

Calculate its surface area : volume ratio. Show your working.

2 cm

_____ [3 marks]

Uses of nanoparticles

1. $PM_{2.5}$ particulates are produced when fuels burn. A scientific study carried out in 2011 estimated that 90 deaths per year are caused by people breathing in $PM_{2.5}$ particulates.

a The study was published in a peer-reviewed scientific journal. How does this affect the validity of the results?

Remember
'Validity' means the suitability of the investigative procedure to answer the question being asked.

_____

_____ [2 marks]

b Some people think that nanoparticles should be banned. Use the information from this question, plus your own knowledge about the benefits and risks of nanoparticles, to evaluate a proposed ban on the use of nanoparticles.

_____

_____

_____

_____

_____ [6 marks]

Metallic bonding

1. Some electrons in a metal are delocalised. What does this mean?

_____

_____ [2 marks]

2. The data in the table below shows the melting points of metals found in Groups 1 and 2 of the periodic table.

Synoptic

a Use the data in the table to compare the melting points of Groups 1 and 2 metals.

Metal	Melting point in °C
Calcium	842
Magnesium	650
Potassium	63
Sodium	98

_____

_____ [1 mark]

b In general, the more delocalised electrons there are in a metal, the stronger its metallic bonds. Use the information to suggest an explanation for your answer to part (a). Use the electronic structure of the metals in your answer.

_____

_____ [4 marks]

Properties of metals and alloys

1. Study the information in the table.

Synoptic

Material	Density in g/cm³	Can it conduct electricity?	Malleability
Iron	7.87	Yes	High
Graphite	2.27	Yes	Low

Explain why each of the materials has its properties. Use what you know about their structure and bonding in your answer.

_____

_____

_____

_____ [6 marks]

Writing formulae

1. A student reacts some calcium carbonate with hydrochloric acid.

Required practical

a Give the formulae for each reactant.

Calcium carbonate: _____

Hydrochloric acid: _____ [2 marks]

Synoptic

b Two of the products of the reaction are water and carbon dioxide. Suggest how the student could measure the volume of carbon dioxide produced.

_____

_____

_____ [1 mark]

Synoptic

c Name the other product of the reaction. _____ [1 mark]

2. The balanced symbol equation for a reaction is

$$C_3H_8 + 5O_2 \rightarrow 3CO_2 + 4H_2O$$

Synoptic

a Identify the type of reaction shown by this equation. _____ [1 mark]

b State the number of

carbon dioxide molecules produced. _____ [1 mark]

oxygen atoms present in the reactants. _____ [1 mark]

Conservation of mass and balanced chemical equations

1. A student mixes solutions of lead(II) nitrate and sodium iodide in a test tube. The balanced symbol equation for the reaction is:

Required practical

$$Pb(NO_3)_2(aq) + 2NaI(aq) \rightarrow PbI_2(s) + 2NaNO_3(aq)$$

a Write this as an ionic equation.

_____ [3 marks]

b They observe that the mixture goes yellow. After a few seconds, the yellow substance sinks to the bottom of the test tube. Name this yellow substance and explain why it sinks to the bottom.

_____

_____ [2 marks]

Maths **c** The student uses 2.4 g of lead nitrate. The total mass at the end of the reaction is 3.9 g. Calculate the mass of sodium iodide used.

_____ [2 marks]

2. A student carried out a neutralisation reaction between calcium hydroxide and hydrochloric acid. Complete the balanced symbol equation for the reaction.

Required practical $Ca(OH)_2 +$ ____$HCl \rightarrow CaCl_2 +$ ____H_2O [2 marks]

> **Remember**
>
> When you balance an equation you can never change the small (subscript) numbers. This changes the substance. For example, O_2 is oxygen and O_3 is the toxic gas ozone.
>
> However, you can change the number of atoms of each substance by adding a number in front of the formula. For example, $2O_2$ contains 4 oxygen atoms.

Mass changes when a reactant or product is a gas

1. A student investigated how the mass changed during the reaction between magnesium carbonate and hydrochloric acid.

The balanced symbol equation for the reaction is:

$MgCO_3(s) + 2HCl(aq) \rightarrow MgCl_2(aq) + H_2O(l) + CO_2(g)$

This is the equipment they used.

The student measured the mass every 30 seconds. They calculated the loss in mass.

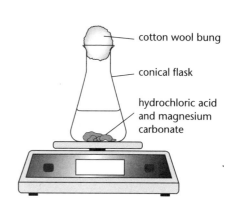

cotton wool bung

conical flask

hydrochloric acid and magnesium carbonate

The table below shows their results.

Time in minutes	Mass loss in g
0	0.00
0.5	0.12
1	0.26
1.5	0.29
2	0.35
2.5	0.40
3	0.42
3.5	0.43
4	0.45
4.5	0.45
5	0.45

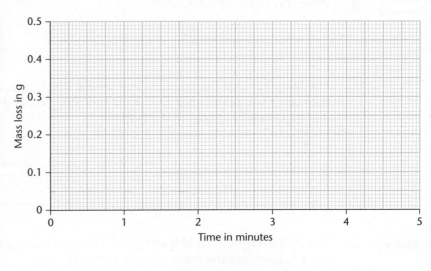

a Plot a graph of these results. [2 marks]

Then draw a line of best fit. [1 mark]

b Explain why there is a loss in mass during the reaction.

_____

_____ [2 marks]

Relative formula mass

1. What is the relative formula mass of carbon dioxide, CO_2? [1 mark]

$A_rC = 12; A_rO = 16$

$M_rCO_2 = 12 + (16 \times 2) = 44$

Marks gained: [1 mark]

Maths

To calculate the relative formula mass (M_r) you add up the relative atomic masses (A_r) of the elements in the formula. If these are not given in the exam question, use your copy of the periodic table.

2. Synoptic

The diagram represents a molecule of paracetamol.

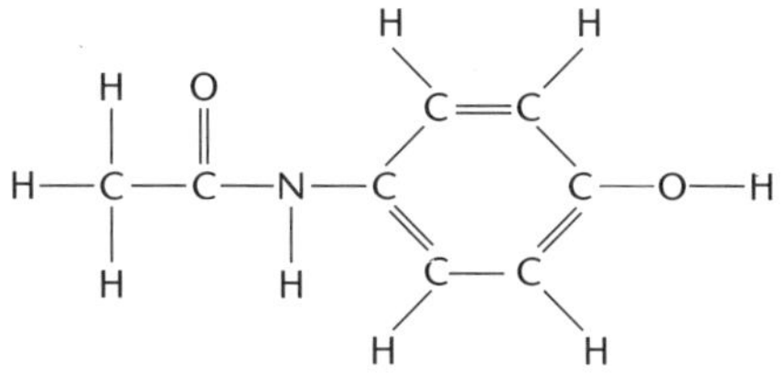

Relative atomic masses (A_r): H = 1; C = 12; N = 14; O = 16

Remember

If a compound contains a metal, the symbol for the metal should go first. The symbols of the non-metals are placed in alphabetical order.

a What is its molecular formula? ____________________ [1 mark]

b Calculate its relative formula mass (M_r).

____ [2 marks]

3. Maths

A student carries out a precipitation reaction.

potassium iodide + lead(II) nitrate → lead(II) iodide + potassium nitrate

$2KI + Pb(NO_3)_2 \rightarrow PbI_2 + 2KNO_3$

(A_r: K = 39; I = 127; Pb = 207; N = 14; O = 16)

Use the symbol equation to show that the relative formula mass of the reactants is equal to the relative formula mass of the products.

____

____

____

____ [6 marks]

Moles

1. Maths Higher Tier only

A sample of a hydrocarbon was burned in excess oxygen. The products were 0.20 mol of water and 6.60 g of carbon dioxide.

a Calculate the number of moles of carbon dioxide produced.

$(A_r: C = 12; O = 16)$

_____ [2 marks]

b Calculate the mass of water produced.

$(A_r: H = 1; O = 16)$

Maths

You can calculate the mass of a substance using the formula:

mass = moles × relative atomic (A_r) or formula mass (M_r)

This can be rearranged to work out the number of moles:

moles = mass ÷ relative atomic (A_r) or formula mass (M_r)

_____ [2 marks]

c There are 6.02×10^{23} atoms, molecules or ions in one mole of a given substance. How many molecules are in 0.2 mol of water? Give your answer to 3 significant figures.

_____ [2 marks]

Amounts of substances in equations

1. A class investigated the change in mass when some magnesium carbonate was heated.

Higher Tier only

The symbol equation for this reaction is:

$MgCO_3 \rightarrow MgO + CO_2$

The investigation followed a typical set of steps.

1 Measure out 4.2 g of magnesium carbonate and put it in a boiling tube.

2 Heat using a Bunsen burner for 5 minutes.

3 Allow to cool and then measure the mass of the solid in the tube.

Synoptic **a** Identify the type of reaction that took place.

_____ [1 mark]

Maths **b** Calculate the maximum mass of magnesium oxide that can be made from 4.2 g of magnesium carbonate. Show your working.

_____ g   [4 marks]

The results worked out for the masses of magnesium oxide from the class were:

2.4 g, 2.2 g, 3.7 g, 2.9 g, 2.1 g, 2.7 g, 3.1 g, 3.2 g

c Suggest why the results are not very accurate.

_____ [1 mark]

d Calculate the percentage uncertainty of these results. Use the formula:

percentage uncertainty = (range ÷ mean) × 100

Show your working. Give your answer to 1 decimal place.

_____ %   [4 marks]

> **Analysing questions**
> Follow these stages:
>
> 1. Calculate the number of moles of magnesium carbonate used.
>
> 2. The symbol equation shows that the number of moles of magnesium carbonate = the number of moles of magnesium oxide.
>
> 3. Calculate the mass of magnesium oxide formed.

Using moles to balance equations

1. 200 g of calcium carbonate is heated to produce carbon dioxide and 112 g of calcium oxide.

Higher Tier only Calculate the number of moles of calcium oxide produced.

Maths _____

_____

_____ [2 marks]

2. The diagram shows the equipment used to react chlorine with iron.

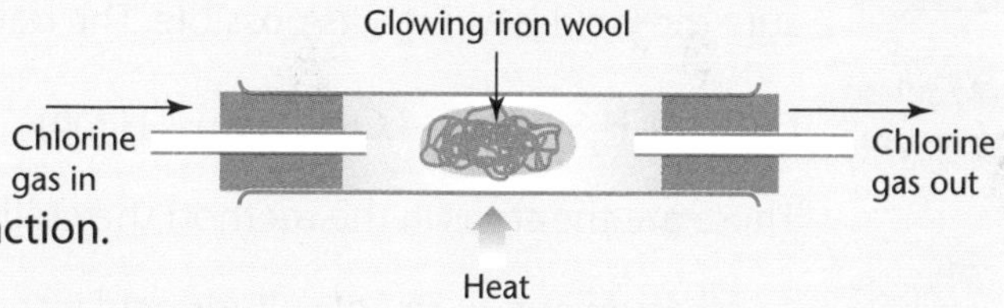

Glowing iron wool

Chlorine gas in

Chlorine gas out

Heat

Synoptic **a** Write the word equation for this reaction.

_____ [1 mark]

b 1.12 g of iron and 2.13 g of chlorine are used. Work out the number of moles of each element used.

(A_r Fe = 56; A_r Cl = 35.5)

_____

_____

_____

_____

_____ [4 marks]

c Use your answers to write the balanced symbol equation for the reaction.

_____ [1 mark]

Limiting reactants

1.

Higher Tier only

Synoptic

The diagram shows a piece of potassium in water at the start (**A**) and at the end of the reaction (**B**).

A   **B**

a Write the balanced symbol equation for this reaction.

_____ [2 marks]

b What is the limiting reactant? Give a reason for your answer.

_____

_____ [2 marks]

2.

Required practical

A student carries out a reaction to make copper sulfate crystals. They are told to make sure the yield is as high as possible. The balanced symbol equation for the reaction is:

$CuO(s) + H_2SO_4(aq) \rightarrow CuSO_4(aq) + H_2O(l)$

These are the steps in the method they follow:

1 Measure out 25 cm³ of sulfuric acid into a beaker.

2 Add one spatula of copper oxide at a time, stirring after each addition.

3 Remove the excess copper oxide.

4 Evaporate the water from the solution.

a Describe how the student could tell when the copper oxide was in excess.

_____ [1 mark]

b Suggest why it is important that the copper oxide is in excess.

_____

_____ [2 marks]

Maths **c** The student used 0.05 mol of sulfuric acid. Calculate the minimum mass of copper oxide needed so that it is in excess. Give your answer to 1 significant figure.

Mass = _____ g [3 marks]

Concentration of solutions

1. Which of the following salt solutions is the most concentrated? Tick **one** box.

Maths

☐ 1 g of salt in 2 cm³ of water.

☐ 5 g of salt in 20 cm³ of water.

☐ 2 g of salt in 10 cm³ of water.

☐ 10 g of salt in 25 cm³ of water.

[1 mark]

Maths

The units of concentration show how to calculate it. If the unit is g/cm³ then you just need to divide the mass of the solute in g by the volume of the solvent in cm³. You might have to convert units, for example from cm³ to dm³.

This formula can be rearranged to give:

mass = volume × concentration; and
volume = mass ÷ concentration.

2. **a** Calculate the concentration in g/cm³ of a solution containing 22 g of sodium chloride in 100 cm³ of water.

Maths

_____

_____

_____ g/cm³ [1 mark]

b Calculate the concentration in g/cm^3 of a solution containing 450 g of copper sulfate in 1.5 dm^3 of water

____

____

______________ g/cm^3 [2 marks]

c Calculate the mass of sodium carbonate needed to make 75 cm^3 of a 20 g/dm^3 solution.

____

______________________________ [2 marks]

3. Higher Tier only

A beaker of salt solution is left uncovered in a classroom. Explain what happens to the concentration of the salt solution over a period of time.

____

____

____

____

____

______________________________ [4 marks]

Using concentrations of solutions in mol/dm^3

1. A scientist needs to make 1.5 dm^3 of a 0.5 mol/dm^3 solution of copper(II) nitrate – $Cu(NO_3)_2$.

Maths

Higher Tier only

a Describe how they would do this. You will need to calculate the mass of copper nitrate needed. Show your working out.

____

____

____

_____

_____

_____ [4 marks]

b The scientist then reacted the copper(II) nitrate solution with dilute sodium hydroxide. The equation for the reaction is:

$Cu(NO_3)_2(aq) + 2NaOH(aq) \rightarrow 2NaNO_3(aq) + Cu(OH)_2(s)$

It took 45.3 cm³ of dilute sodium hydroxide to completely react with 20 cm³ of the 0.5 mol/dm³ copper nitrate solution. Calculate the concentration of the sodium hydroxide solution. Give your answer to 1 decimal place.

_____

_____

_____

_____

_____ [4 marks]

Amounts of substances in volumes of gases

1.

Higher Tier only

Maths

Calculate the volume of 220 g of carbon dioxide at room temperature and pressure.

Remember
At room temperature and pressure (20 °C and 1 atm) the volume of 1 mole of **any** gas is 24 dm³.

_____

_____ [2 marks]

2. The diagram shows the equipment used to measure the change in mass as powdered zinc reacts with hydrochloric acid. The symbol equation for the reaction is:

$Zn + 2HCl \rightarrow ZnCl_2 + H_2$

a Predict what will happen to the mass recorded on the scales as the reaction takes place. Give a reason for your answer.

_____

_____ [2 marks]

Maths **b** Calculate the volume of hydrogen produced when 2.0 g of zinc has reacted.

A_r : Zn = 65

_____

_____ [2 marks]

3. 3.2 g of a hydrocarbon gas has a volume of 4800 cm³.

Maths Work out whether the gas is methane (CH_4), ethane (C_2H_6) or propane (C_3H_8). (A_r : C = 12; H = 1)

Show how you arrive at your answer.

The gas is _____ [4 marks]

Percentage yield

1. **a** 3.6 g of magnesium is burned. The symbol equation for the reaction is:

Higher Tier only

$$2Mg + O_2 \rightarrow 2MgO$$

Remember

To calculate a percentage yield you use this formula:

$$\% \text{ yield} = \frac{\text{mass of product actually produced}}{\text{maximum theoretical mass of product}} \times 100$$

Calculate the maximum theoretical mass of magnesium oxide produced. [3 marks]

Number of moles = mass ÷ A_r

Moles of magnesium = 3.6 ÷ 24 = 0.15 [1 mark]

The symbol equation shows us that moles of Mg = moles of MgO

M_r MgO = 24 + 16 = 40; [1 mark] mass = moles × M_r; mass of MgO = 0.15 × 40 = 6 g [1 mark]

Marks gained: [3 marks]

b The actual mass of magnesium oxide produced is 4.2 g. Calculate the percentage yield.

Percentage yield = (4.2 ÷ 6) × 100 = 70%

Marks gained: [1 mark]

2. A scientist prepares a pure, dry salt. These are the method steps she uses:

Required practical

1 Pour 25 cm³ of 0.5 mol/dm³ ethanoic acid into a conical flask. Add an excess of magnesium.

2 Wait till the reaction stops and then filter off the remaining magnesium.

3 Pour the filtrate into an evaporating dish and leave it in a warm place.

The equation for the reaction is:

ethanoic acid + magnesium → magnesium ethanoate + hydrogen

$2CH_3COOH(aq) + Mg(s) \rightarrow (CH_3COO)_2Mg(aq) + H_2(g)$

Synoptic **a** Describe how the scientist will know when they have added enough magnesium.

_____ [1 mark]

Higher Tier only **b** Calculate the maximum theoretical mass of magnesium ethanoate produced.

Synoptic _____

Maths _____

_____

_____ [4 marks]

Atom economy

> **Remember**
>
> Atom economy is shown as a percentage and is calculated using this formula:
>
> $$\frac{\text{relative formula mass of desired product from equation}}{\text{sum of relative formula masses of all reactants from equation}} \times 100$$

1. Why do companies choose reactions with a high atom economy? Tick **two** boxes.

☐ It is important for sustainable development.

☐ To increase the amount of reactants they use.

☐ To increase the amount of money they make.

☐ To be able to calculate yields. [2 marks]

2. Drug companies make the useful drug aspirin. Aspirin can be made by reacting salicylic acid with ethanoic anhydride. The equation for this reaction is:

Maths

$$C_7H_6O_3 \quad + \quad C_4H_6O_3 \quad \rightarrow C_9H_8O_4 + \quad CH_3COOH$$

salicylic acid + ethanoic anhydride → aspirin + ethanoic acid

a Calculate the atom economy for this reaction. Show your working.

_____

_____

_____

_____ [3 marks]

Higher Tier only

b Aspirin can also be made in another reaction:

$$C_7H_6O_3 + CH_3COOH \rightarrow C_9H_8O_4 + H_2O$$

salicylic acid + ethanoic acid → aspirin + water

This reaction gives a higher atom economy than the first but it is rarely used. Suggest one reason why.

_____

_____ [1 mark]

Metal oxides

1. For each of these reactions, say if it is an example of reduction or oxidation.

a magnesium + oxygen → magnesium oxide _____ [1 mark]

b mercury oxide → mercury + oxygen _____ [1 mark]

c $2CO + O_2 \rightarrow 2CO_2$ _____ [1 mark]

2. Zinc oxide is added to a wide range of products including paints, ceramics and plastics.

a Suggest how zinc oxide is made from pure zinc in industry. Write a balanced symbol equation for this reaction. (Zinc ions have a charge of +2.)

_____

_____ [4 marks]

Synoptic **b** 300 kg of zinc is used to make some zinc oxide.
The maximum theoretical yield of zinc oxide is 373.8 kg.
Explain why the mass of the product is higher than that of the reactant.

_____ [1 mark]

> **Synoptic**
> In this question you need to link your knowledge from more than one area of chemistry. There will be 'synoptic' questions like this in the exam.

Synoptic **c** The actual mass of zinc oxide produced is 350 kg.
Calculate the percentage yield.

Maths

_____

_____ [1 mark]

Reactivity series

1. Which statement explains why sodium is a more reactive metal than lithium?
Tick **one** box.

☐ Only sodium forms positive ions. ☐ Sodium has more electrons.

☐ Sodium loses electrons more easily. ☐ Sodium has a lower melting point. [1 mark]

2. Use the reactivity series below to answer this question.

Most reactive

Potassium
Sodium
Calcium
Magnesium
Aluminium
Carbon
Zinc
Iron
Tin
Lead
Hydrogen
Copper
Silver
Gold
Platinum

Least reactive

Required practical

A group of students had four different grey metals. They knew that the metals were calcium, magnesium, zinc and silver but did not know which metal was which.

The teacher asked the class to plan a method that could be used to identify each metal. They were given some dilute hydrochloric acid and some water.

a Describe a method the students could use to identify the metals. You should include the results of the tests you describe.

_____

_____

_____

_____

_____

_____ [6 marks]

b Explain why the students were not given sodium to identify.

_____

_____ [2 marks]

> **Remember**
> Less reactive metals do not react with water.

Reactivity series – displacement

Use the reactivity series to answer these questions.

1. Which metal listed in the reactivity series can be used to displace lead from lead sulfate?

Tick **one** box.

☐ Copper ☐ Iron

☐ Gold ☐ Lead [1 mark]

2. A scientist has two grey metals. They know that one is zinc and the other is platinum. Explain how they could identify the metals using copper sulfate solution.

_____

_____ [3 marks]

Extraction of metals

1. Most metals are extracted from compounds using chemical reactions.

a Draw **one** line from each metal to the method of its extraction.

Remember
Only metals less reactive than carbon can be extracted using reduction with carbon.

Metal	Method of extraction
iron	
Magnesium	reduction with carbon
Sodium	electrolysis
Lead	

[4 marks]

b Explain why platinum does not need to be extracted from ores.

_____

_____ [3 marks]

2. Some metals can be extracted by heating their ores with carbon. An equation for one example of this reaction is:

$$ZnO_2 + C \rightarrow Zn + CO_2$$

a Name the substance that is reduced. _____ [1 mark]

b Name the substance that is oxidised. _____ [1 mark]

Synoptic **c** Describe one environmental consequence associated with extracting metals in this way.

_____

_____ [2 marks]

Command words
You are asked to 'describe' an issue. This means that you will need to *discuss* the consequence, rather than just name it.

Oxidation and reduction in terms of electrons

1. Which statement correctly describes what happens during reduction? Tick **one** box.

Higher Tier only

☐ A metal atom gains electrons to form ions.

☐ A metal atom loses electrons to form ions.

☐ A non-metal atom gains electrons to form ions.

☐ A non-metal atom loses electrons to form ions. [1 mark]

Remember
OILRIG: **O**xidation **I**s electron **L**oss; **R**eduction **I**s electron **G**ain.

2. A student carried out a reaction between copper sulfate and iron. The symbol equation for the reaction is:

$Fe(s) + CuSO_4(aq) \rightarrow FeSO_4(aq) + Cu(s)$

a Write the ionic equation for the reaction. [2 marks]

_____ [2 marks]

b Identify what was:

Reduced _____

Oxidised _____

during the reaction. [2 marks]

3. Some cars run on electricity produced by a hydrogen–oxygen fuel cell. The half equations below take place inside the fuel cell:

A: $H_2(g) - 2e^- \rightarrow 2H^+(aq)$

B: $4H^+(aq) + O_2(g) + 4e^- \rightarrow 2H_2O(g)$

Does **B** show oxidation or reduction? Give a reason for your answer.

_____ [2 marks]

Reactions of acids with metals

1. What are the products made when zinc reacts with sulfuric acid? Tick **two** boxes.

☐ Hydrogen   ☐ Water

☐ Zinc sulfide   ☐ Zinc sulfate   [2 marks]

2. A student adds some magnesium to some hydrochloric acid.

Complete the balanced symbol equation for the reaction.

$Mg + \_\_\_\_ HCl \rightarrow _____ + H_2$ [2 marks]

3. Iron reacts slowly with sulfuric acid.

a Name the salt produced in this reaction. _____   [1 mark]

Higher Tier only   **b** State what is:

Reduced _____

Oxidised _____   [2 marks]

Neutralisation of acids and making salts

1. Which statements about calcium oxide are correct? Tick **two** boxes.

☐ It is soluble in water.   ☐ It is insoluble in water.

☐ It neutralises alkalis.   ☐ It neutralises acids.   [2 marks]

2. A student carried out a neutralisation reaction. The reactants they used were in bottles labelled 'KOH' and 'HCl'.

a Name the acid and the alkali they used.

_____   [2 marks]

b Complete the symbol equation for the reaction.

$KOH + HCl \rightarrow _____ + H_2O$ [1 mark]

3. A student reacts some copper carbonate with some sulfuric acid. The diagram shows the equipment they use.

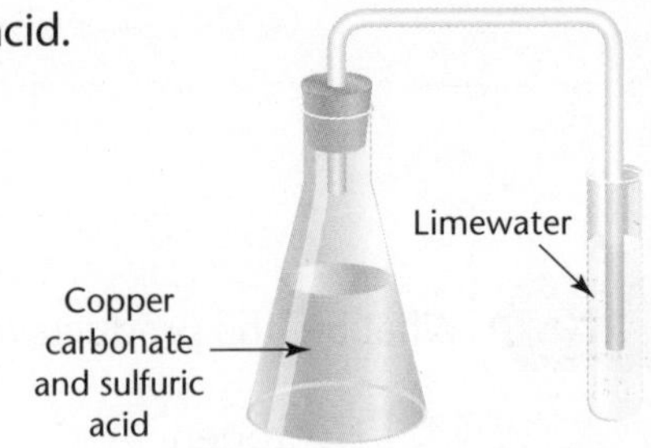

Limewater

a Complete the symbol equation for this reaction. The charges on the ions are Cu^{2+}, CO_3^{2-} and SO_4^{2-}.

Copper carbonate and sulfuric acid

$CuCO_3 + H_2SO_4 \rightarrow$ _____ +

_____ + _____

[3 marks]

Synoptic **b** Predict what will happen to the limewater. Explain why.

_____ [2 marks]

Making soluble salts

1. Salts can be made by reacting a metal with an acid.

a Name the metal and the acid you would use to make the salt called zinc chloride.

Metal: _____ Acid: _____ [2 marks]

Synoptic **b** Describe a test that a student could carry out to prove that hydrogen is made during this reaction.

_____

_____ [2 marks]

> **Synoptic**
> For this question, you will need to apply your knowledge of tests for gases from Section 8.

2. Describe a method for making copper sulfate crystals using copper oxide and dilute sulfuric acid. You should include:

Required practical

- the names of the pieces of equipment used

- the purpose of each step

- safety advice.

_____

_____

_____

_____

_____

_____

_____

_____

_____

_____ [6 marks]

pH and neutralisation

1. Draw links from the pH range to the type of solution described.

pH range	Type of solution
7	acidic
8–14	alkaline
1–6	neutral

[3 marks]

2. Complete the equation to show the ionic equation for the neutralisation reaction.

$H^+(aq) +$ _____ $(aq) \rightarrow$ _____ (l) [2 marks]

3. A solution of sodium chloride can be made by adding hydrochloric acid to sodium hydroxide.

Required practical **a** Explain how you could use universal indicator to check when the reaction is *just* complete.

_____

_____

_____ [3 marks]

b State **one** disadvantage of using universal indicator for this technique. Suggest an alternative that does not have this problem.

_____

_____ [2 marks]

Titrations

1. Some students carried out a titration between sodium hydroxide and 0.100 mol/dm³ sulfuric acid. The diagram shows the equipment they used.

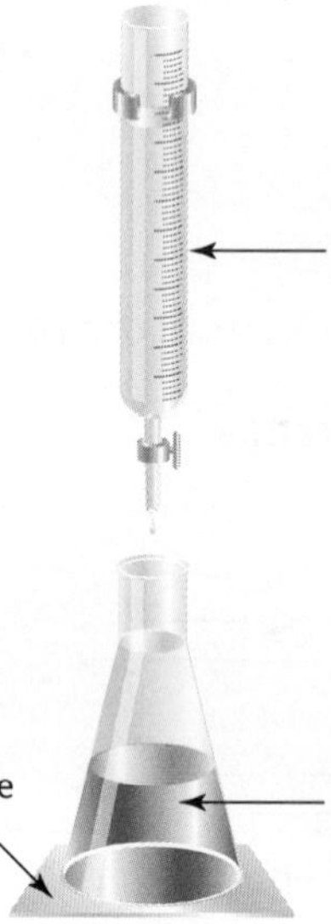

Required practical

a Describe how to use the equipment to obtain an accurate value for the volume of acid needed to neutralise the sodium hydroxide.

← 0.100 mol/dm³ sulfuric acid

_____

_____

_____

_____

_____

_____ [4 marks]

White tile

25 cm³ sodium hydroxide with a few drops of indicator

b Suggest the function of the white tile.

_____

_____ [2 marks]

2. One student found that 25.0 cm³ of sodium hydroxide solution was neutralised by exactly 22.30 cm³ of 0.100 mol/dm³ sulfuric acid solution.

Higher Tier only

The balanced symbol equation for the reaction is:

Maths

$H_2SO_4(aq) + 2NaOH(aq) \rightarrow Na_2SO_4(aq) + 2H_2O(l)$

Calculate the concentration of the sodium hydroxide solution in mol/dm³. Give your answer to 3 decimal places.

_____

_____

_____

_____

_____

_____

Concentration of sodium hydroxide solution _____ mol/dm³.   [6 marks]

Analysing questions

1. Calculate the number of moles of sulfuric acid using moles = concentration × volume. Make sure you convert all volumes to dm³.

2. Use the balanced equation to work out the ratio of moles of acid : moles of alkali.

3. Calculate the concentration of the alkali using concentration = moles ÷ volume.

Strong and weak acids

1.   Here are three bottles of different acids.

Higher Tier only

a   Which acid is the weakest?

_____   [1 mark]

1 mol/dm³
Nitric acid

Acid A

2 mol/dm³
Citric acid

Acid B

0.1 mol/dm³
Hydrochloric acid

Acid C

b   What ions are formed when hydrochloric acid ionises? Tick **2** boxes.

☐ H^+ ☐ H^- ☐ Cl^+ ☐ Cl^- [2 marks]

2.   Look at the information in the table.

Concentration of hydrogen ions in mol/dm³	pH
10^{-1}	1
10^{-2}	2
10^{-3}	3
10^{-4}	4

a What is the pH of a solution with a concentration of 0.000001 mol/dm^3 of hydrogen ions.

_____ [2 marks]

b A scientist dilutes an acid with pH 2 by adding some water to it. Predict what happens to the pH of the acid. Give a reason for your answer.

_____

_____ [2 marks]

Command words

If you are asked to 'predict' something, write what you think the outcome will be. You may also be asked to give your reasoning – why you think this will happen – using scientific ideas.

3. Explain why sulfuric acid is classed as a strong acid. You should include information about ionisation in your answer.

_____

_____ [2 marks]

The process of electrolysis

1. During electrolysis why does the electrolyte have to be molten or in solution?

_____ [1 mark]

2. A student carries out electrolysis of a solution of copper chloride ($CuCl_2$).

a Complete the diagram to show a labelled diagram of the equipment they use. Name the electrodes. [4 marks]

b Copper ions are present in the copper chloride solution. Predict which electrode they move towards. Give a reason for your answer.

_____

_____ [2 marks]

3. Electrons are transferred at the electrodes. Some ions gain electrons at the negative electrode.

Higher Tier only

a Complete the half equation to show this.

$X^+ +$ _____ $\rightarrow$ _____ [2 marks]

b Is this oxidation or reduction? _____ [1 mark]

Electrolysis of molten ionic compounds

1. Which compounds will conduct electricity when molten?
Tick **two** boxes.

☐ Sodium chloride (NaCl)

☐ Silicon dioxide (SiO_2)

☐ Calcium oxide (CaO)

☐ Glucose ($C_6H_{12}O_6$)

> **Remember**
> Only ionic compounds will conduct electricity when molten. Ionic compounds are made up of metal ions bonded to non-metal ions.

[2 marks]

2. Look at this diagram.

a As the lead bromide is heated it starts to melt. Explain why the bulb lights.

_____

_____

_____ [3 marks]

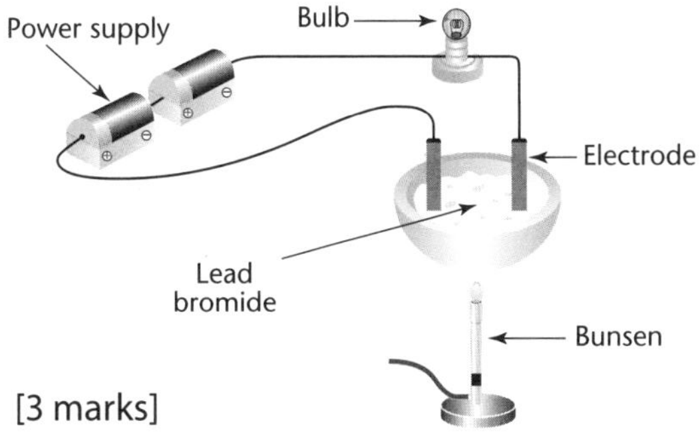

Power supply   Bulb

Electrode

Lead bromide

Bunsen

b Predict what will be seen at each electrode. Give a reason for your answer.

_____

_____

_____

_____ [4 marks]

1. The diagram shows how aluminium is extracted from aluminium oxide using electrolysis.

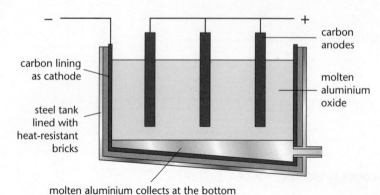

Synoptic **a** Explain why aluminium cannot be extracted by heating aluminium oxide with carbon.

_____ [1 mark]

b Aluminium collects at the bottom of the tank. Explain why.

_____

_____ [2 marks]

Higher Tier only **c** Complete the half equation for the formation of the metal.

$Al^{3+} +$ _____ $e^- \rightarrow Al$ [1 mark]

d A chemical reaction takes place at the anodes. Explain why this means they need to be replaced regularly. Include a balanced symbol equation in your answer.

_____

_____

_____

_____ [4 marks]

Electrolysis of aqueous solutions

1. A student carried out electrolysis on copper sulfate ($CuSO_4$) solution.

Required practical **a** Which ions are attracted to the anode? Tick **one** box.

[ ] Copper ions and hydrogen ions.

[ ] Oxygen ions and sulfur ions.

[ ] Hydrogen ions and hydroxide ions.

[ ] Sulfate ions and hydroxide ions. [1 mark]

> **Remember**
> Solutions contain water. Water can split into hydrogen (H^+) and hydroxide (OH^-) ions.

b Bubbles are produced at the anode. The student predicts that the bubbles contain oxygen. Describe how they could test this hypothesis.

_____

_____ [2 marks]

Higher Tier only **c** The student was expecting bubbles of hydrogen gas to be produced at the cathode. Instead they observe that it is covered in a brown substance. Suggest a reason for this. Include a half equation for this process.

_____

_____

_____ [4 marks]

2. Predict the products discharged at the electrodes when electrolysis is carried out on sodium bromide solution.

Anode: _____ Cathode: _____ [2 marks]

Half equations at electrodes

1. **A**, **B** and **C** are three different half equations.

Higher Tier only **A**: $2Br^- \rightarrow Br_2 + 2e^-$ **B**: $2H^+ + 2e^- \rightarrow H_2$ **C**: $Zn^{2+} + 2e^- \rightarrow Zn$

Tick the correct boxes in these questions.

a Which half equation happens at the anode? [1 mark]

A ☐   B ☐   C ☐

b Which half equation shows oxidation? [1 mark]

A ☐   B ☐   C ☐

c Which half equation shows the reduction of a non-metal ion? [1 mark]

A ☐   B ☐   C ☐

2.

Higher Tier only

The electrolysis of sodium chloride (NaCl) solution is carried out on a large scale in industry. This diagram shows the equipment used.

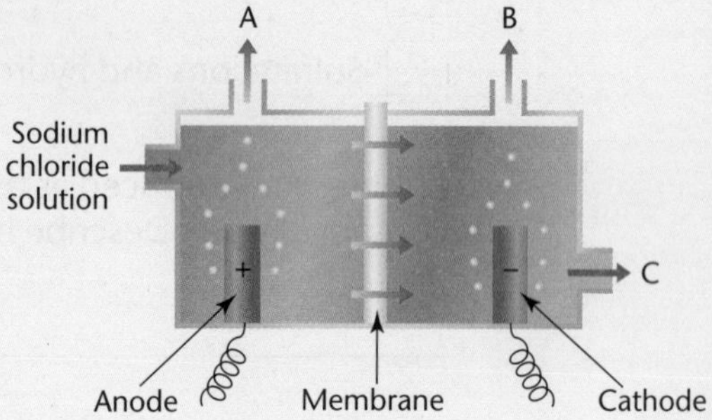

a Chlorine gas is produced at the anode (**A**). Complete the half equation for the process at the anode.

$2Cl^- - \underline{\hphantom{xxx}} \rightarrow Cl_2$ [1 mark]

b Hydrogen is produced at the cathode (**B**). Write the half equation for this process.

_____ [2 marks]

c Suggest the name of the final product (**C**). Give a reason for your answer.

_____

_____ [2 marks]

3.

Higher Tier only

Complete these half equations for the electrolysis of copper sulfate ($CuSO_4$) solution.

a At the cathode: $Cu^{2+} + \underline{\hphantom{xxx}} \rightarrow Cu$

b At the anode: $\underline{\hphantom{xxx}} - 4e^- \rightarrow O_2 + 2H_2O$ [2 marks]

Exothermic and endothermic reactions

1. Draw **one** line from each practical application to the type of reaction.

Sports injury pack

Hand warmer

Self-heating can

Photosynthesis

Exothermic reaction

Endothermic reaction

[3 marks]

2. Abi and Heather mixed some citric acid with 25 cm³ of sodium hydrogen carbonate solution. They saw fizzing and the test tube got cooler.

Required practical

a Name the gas that was produced.

_____ [1 mark]

b Identify the type of reaction that produces a decrease in temperature.

_____ [1 mark]

Maths **c** The girls recorded the following data:

Starting temperature = 19 °C.

Lowest temperature = 5 °C.

Calculate the change in temperature.

Change in temperature = _____ °C   [1 mark]

3. Jack wants to investigate the temperature changes that take place when magnesium metal is added to dilute hydrochloric acid. During the investigation, he uses this apparatus.

Required practical

Jack starts by making this hypothesis: 'The bigger the piece of magnesium used, the smaller the temperature change when reacting with hydrochloric acid.'

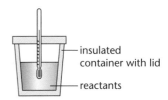

insulated container with lid

reactants

a Complete the equation, balance it and identify the missing state symbols.

$Mg(s) + \underline{\quad} HCl(aq) \rightarrow MgCl_2 \underline{\quad} + H_2 \underline{\quad}$ [3 marks]

b Suggest two variables that Jack should keep the same.

_____

_____ [2 marks]

Jack's preliminary results are shown in the table

Amount of Mg in cm	Temperature °C		
	Starting	Final	Change
1	22.0	22.8	
2	22.0	23.6	
3	21.5	23.0	
4	22.0	25.2	
5	19.5	23.5	

c Write the temperature changes in the table. [2 marks]

d Study the data in the table. Suggest what Jack can conclude from his preliminary results.

_____ [1 mark]

Reaction profiles

1. This question is about the reaction between hydrogen and oxygen. The equation for the reaction is:

$2H_2(g) + O_2(g) \rightarrow 2H_2O(g)$

a Complete the reaction profile shown here. Draw labelled arrows to show:

- The energy given out during the reaction.

- The activation energy. [3 marks]

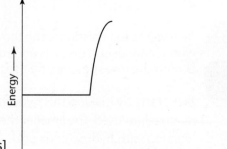

b Explain why hydrogen does not spontaneously combust when it comes into contact with oxygen at room temperature.

_____

_____ [2 marks]

c Describe and explain what would happen if a naked flame came in contact with the mixture.

_____

_____ [3 marks]

2. Hydrogen bromide decomposes to form hydrogen and bromine:

$2HBr(g) \rightarrow H_2(g) + Br_2(g)$

The reaction is endothermic. Complete and label the reaction profile for this reaction, identifying the activation energy. [3 marks]

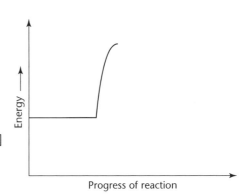

Energy change of reactions

•••

1. Hydrogen bromide decomposes to form hydrogen and bromine. The table lists some bond energies. [3 marks]

Higher Tier only

Maths

Bond	Bond energy in kJ/mol
H–Br	366
H–H	436
Br–Br	193

Use the bond energies to calculate the overall energy change for the reaction:

$2HBr(g) \rightarrow H_2(g) + Br_2(g)$

First, work out the bond energies of the reactants:

 2 × H–Br = 2 × 366 = 732 kJ/mol. [1 mark]

Next, work out the bond energies of the products:

H–H = 436 kJ/mol and Br–Br = 193 kJ/mol.

Total = 436 + 193 = 629 kJ/mol. [1 mark]

Now work out the difference between the bond energies of the reactants and products:

732 – 629 = 103 kJ/mol. [1 mark]

Less energy is released than is taken in, so the reaction is endothermic.

Marks gained: [3 marks]

2.

Higher Tier only

Explain what happens in terms of bond breaking and bond making during exothermic and endothermic reactions.

Literacy

In this question you will be assessed on using good English, organising information clearly and using specialist terms where appropriate. Use your knowledge and understanding of chemical reactions and energy changes.

_____

_____

_____

_____

_____

_____ [6 marks]

Cells and batteries

1. The diagram shows a simple electrochemical cell.

a Suggest a suitable electrolyte for the reaction.

_____ [1 mark]

b Explain why a voltage is produced.

_____

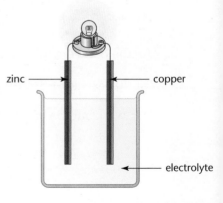

zinc ← → copper

electrolyte

_____ [3 marks]

Higher Tier only

c The following reactions take place in the cell:

$$Zn \rightarrow Zn^{2+} + 2e^-$$

$$Cu^{2+} + 2e^- \rightarrow Cu$$

Identify which metal is oxidised and write the ionic equation for the overall reaction.

_____

_____ [2 marks]

The potential produced by the cell is 0.7 V. Use the electrochemical series below (reactivity decreasing from left to right) to answer the following questions.

$$Mg > Zn > Fe > Pb > Cu > Ag$$

d Predict what happens to the cell voltage when the zinc in the diagram is replaced by lead.

_____ [1 mark]

e Predict what happens to the cell voltage when the copper is replaced by silver.

_____ [1 mark]

f Explain why the cell does not produce a voltage when both electrodes are iron.

_____ [1 mark]

2. David needs a new battery for his bicycle torch. When he goes to buy a new one he discovers that a non-rechargeable alkaline battery will cost him 99p but a rechargeable one costs £2.50.

Compare the two types of batteries to help decide which type to buy.

_____

_____

_____

Command words

The command word 'compare' means to describe both differences *and* similarities between things.

_____ [4 marks]

Fuel cells

• •

1. Hydrogen can be used as a fuel.

a Explain why hydrogen is a better fuel for the environment than fossil fuels such as coal or petrol.

_____

_____

_____

_____ **[4 marks]**

Higher Tier only

b The overall reaction in a hydrogen fuel cell involves the oxidation of hydrogen to produce water. Complete the half equations for each electrode reaction.

Remember
The numbers of electrons in the balanced half equation cancel each other out.

At the negative electrode: $2H_2 \rightarrow$ ___ $H^+ +$ ___ e^-

At the positive electrode: $4H^+ + O_2 +$ ___ $e^- \rightarrow$ ___ H_2O **[2 marks]**

c The diagram shows the negative and positive electrodes of a fuel cell.

Explain why an electric current is produced in the external circuit.

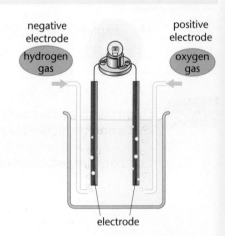

negative electrode   positive electrode

hydrogen gas   oxygen gas

electrode

_____

_____

_____

_____ **[3 marks]**

Measuring rates of reaction

Required practical

1. Calcium carbonate reacts with dilute hydrochloric acid:

$$CaCO_3 + 2HCl \rightarrow CaCl_2 + H_2O + CO_2$$

A student investigated the rate of mass loss in this reaction. This is the method they used:

1 Put a known mass of calcium carbonate into a conical flask.

2 Measure out 10 cm³ of dilute hydrochloric acid using a measuring cylinder.

3 Place the flask on a balance, pour in the acid and put a cotton wool bung in the top of the flask.

4 Record the mass of the flask and its contents every 15 seconds.

The student set up the apparatus as shown opposite.

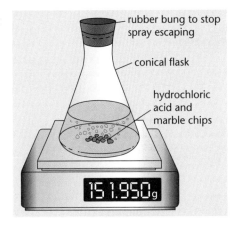

rubber bung to stop spray escaping

conical flask

hydrochloric acid and marble chips

151.950g

a Identify the error in the way the apparatus was set up. Describe what would happen if this apparatus were used.

_____

_____ [2 marks]

Maths   **b** Once the error had been rectified, some results were recorded.

Time in seconds	Mass of flask in grams
0	80.906
15	80.774
30	80.757
45	80.570
60	80.744
75	80.733
90	80.721
105	80.716
130	80.699
145	80.686
160	80.686

Identify the anomalous result in this set. Suggest what may have happened to cause this.

_____ [2 marks]

Maths **c** Calculate the mass of the carbon dioxide produced during the experiment.

_____ [1 mark]

Higher Tier only **d** Calculate the number of moles of CO_2 gas produced. (A_r : C = 12; O = 16)

Maths

Synoptic

_____

_____ [2 marks]

Maths

Remember that molar mass is the relative formula mass in grams. You will also need to recall the equation that links molar mass, mass and moles.

Calculating rates of reaction

1.

Required practical

Maths

Richard and Mark were investigating the rate of decomposition of hydrogen peroxide. When they added 1.0 g of manganese oxide to 100 cm³ of hydrogen peroxide, they collected 60 cm³ of gas in 7 minutes.

Maths

You will need to recall the equation:

$$\text{mean rate of reaction} = \frac{\text{quantity of product formed}}{\text{time taken}}.$$

Don't forget to work out the correct units.

a Calculate the mean rate of the reaction. Give your answer to 3 significant figures.

_____ [3 marks]

b The results are recorded in the table.

Time in minutes	0	1	2	3	4	5	6	7	8
Volume in cm³	1	20	33	44	52	58	59	60	60

Plot a graph of volume of oxygen against time.

[5 marks]

Higher Tier only **c** Use the graph to work out the rate of reaction after 1.5 minutes.

_____ [4 marks]

Effect of concentration and pressure

1. The diagram shows a gas syringe containing particles of two different reacting gases.

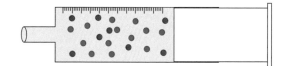

a As the syringe plunger is pushed in, predict what will happen to the rate of the reaction between the gases. Give a reason for your answer.

_____

_____ [2 marks]

Maths **b** A typical collision frequency for gas particles is 5×10^{10} per second. Calculate the time taken for 1×10^5 collisions to occur. Write your answer in standard form.

_____ [2 marks]

c Calculate how many collisions would take place in 2.5 seconds. Write your answer in standard form.

_____ [2 marks]

2. This graph shows the total volume of hydrogen produced when some magnesium ribbon reacts with excess hydrochloric acid.

a Sketch on the graph the result you would expect to obtain if the same mass of magnesium were treated with:

- more concentrated acid [2 marks]

- very dilute acid, so the magnesium is now in excess. [2 marks]

> **Maths**
> 'Collision frequency' is the number of successful collisions per second. Use this definition to deduce an expression for the time taken. Don't forget the rules of division when working in standard form.

b Use your understanding of collision theory to explain your answers to part (a).

_____

_____

_____

_____

_____ [5 marks]

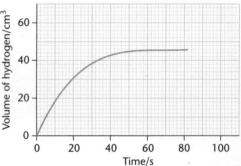

Effect of surface area

1. Susie was investigating how surface area affects the rate of a reaction. She mixed a known amount of calcium carbonate (marble chips) with a known volume of hydrochloric acid and recorded the volume of gas produced every 10 s. She then repeated the experiment using different-sized marble chips.

Maths

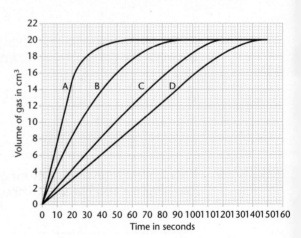

The diagram shows the results of four experiments (A, B, C and D).

a Identify the experiment that used the largest marble chips.

_____ [1 mark]

b Calculate the rate of reaction observed near the start of experiment A. Compare it with the rate of reaction observed during experiment C.

_____

_____ [3 marks]

c Using your knowledge of particle theory, explain your answer to part (b).

_____

_____

_____ [4 marks]

Higher Tier only

Synoptic

d Calculate the number of moles of gas produced during the reaction, given that 1 mole of a gas occupies 24 dm³. Write your answer in standard form.

_____

_____ [3 marks]

Effect of temperature

1.

Required practical

In the reaction between sodium thiosulfate and hydrochloric acid, a precipitate of sulfur is produced slowly. The equation for the reaction is:

$Na_2S_2O_3(aq) + 2HCl(aq) \rightarrow 2NaCl(aq) + H_2O(l) + SO_2(g) + S(s)$

Describe how you would design an experiment to investigate how temperature affects the rate of reaction.

_____

_____

_____

_____

_____

_____

Literacy

In 6-mark questions like this you will be assessed on using good English, organising information clearly and using specialist terms where appropriate. Use the information you have been given and your knowledge and understanding of practical chemistry to answer it.

_____

_____   [6 marks]

2. A student investigated the effect of temperature on the reaction between sodium thiosulfate and dilute hydrochloric acid. The results are recorded in the table.

Practical

Maths

Experiment number	Temperature in °C	Time for X to disappear in s	Time for X to disappear in s	Mean time in s	Rate in s^{-1}
1	40	25	24	24.5	0.041
2	50	20	21	20.5	0.049
3	60	16	16	16.0	0.063
4	70	11	12		
5	80	10	10		

a   Complete the table.   [2 marks]

b   Draw a graph of temperature (*x*-axis) against rate (*y*-axis).   [5 marks]

c   Use your graph to draw a conclusion about the change of rate of reaction with temperature.

_____[1 mark]

Effect of a catalyst

1. Complete the sentence to provide a definition for a catalyst:

Catalysts change the _____ of a reaction, but are _____ [2 marks]

2. The diagram shows a reaction profile.

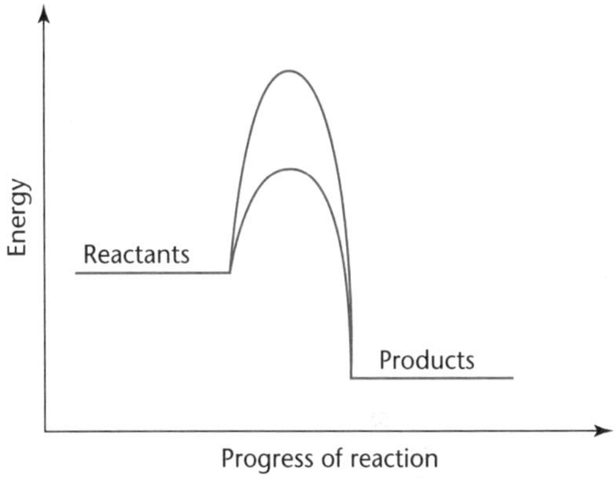

Synoptic **a** Draw arrows on the profile to show:

- the activation energy for the reaction without a catalyst present. [1 mark]

- the activation energy for the reaction with a catalyst present. [1 mark]

b Explain how a catalyst works.

_____

_____ [2 marks]

3. A group of scientists (A) measured the activation energy for a reaction and found it to be 69 kJ/mol. A different group (B) repeated the measurement and found it to be 49 kJ/mol. Suggest a reason for the difference in results.

_____

_____ [2 marks]

4. Many industrial processes use a catalyst. For example, zeolites are used in the cracking of long-chain hydrocarbons. Suggest why such processes use a catalyst.

_____

_____

_____

_____ [4 marks]

Reversible reactions and energy change

1. Methanol is an important industrial alcohol that is manufactured directly by the reaction of hydrogen with carbon monoxide. Complete the equation by balancing it.

$$CO(g) + \underline{\hspace{1cm}} H_2(g) \rightleftharpoons CH_3OH(g)$$

[1 mark]

2. Heating blue copper sulfate, $CuSO_4 \cdot 5H_2O$, and then adding water to it is an example of a reversible reaction. The reaction is endothermic in the forward reaction and exothermic in the reverse reaction.

blue $\rightleftharpoons$ white
hydrated   anhydrous

> **Analysing questions**
> You have been given the formula for hydrated blue copper sulfate. You need to make sure that there is the same number of atoms on both sides of the equation.

Higher Tier only

a Write a balanced equation for the reaction; include state symbols.

_____

b Explain why the forward reaction is endothermic.

_____ [2 marks]

c Describe the conditions required for the reverse reaction to take place.

_____ [1 mark]

d Describe what is observed during the reverse reaction.

_____ [2 marks]

e The energy change for the forward reaction is 78.22 kJ/mol. Predict the energy change for the reverse reaction. Give a reason for your answer.

_____

_____ [2 marks]

Equilibrium and Le Chatelier's principle

1. Describe the conditions needed for a chemical equilibrium to be reached.

_____

_____ [2 marks]

2. The reaction $W + X \rightleftharpoons Y + Z$ is at equilibrium. The concentrations of W and X are higher than those of Y and Z. Identify where the position of equilibrium lies.

_____ [1 mark]

3. In a fizzy drink this equilibrium is in place:

$$CO_2(g) \rightleftharpoons CO_2(aq)$$

   a Identify where the position of equilibrium lies when the bottle is closed.

   _____ [1 mark]

   b Explain what happens to the equilibrium when the bottle is left open.

   _____ [2 marks]

4. State what Le Chatelier's principle can be used to predict.

Higher Tier only

_____ [1 mark]

5. A saturated solution of sodium chloride is at equilibrium. Describe what is happening at the particle level. Include a balanced or ionic equation in your answer.

Higher Tier only

Synoptic

_____

_____

_____

_____ [4 marks]

Changing the position of equilibrium

1.

Higher Tier only

Synoptic

Sulfuric acid is made in the Contact process. In one stage, sulfur dioxide is converted to sulfur trioxide in an exothermic reaction.

a Complete the symbol equation for the reaction by balancing it. [2 marks]

___SO_2(g) + O_2(g) ⇌ ___SO_3(g)

b Which pressure would you choose to favour the forward reaction: 1 atm, 100 atm or 200 atm? Give a reason for your answer.

_____

_____

_____ [3 marks]

c The Contact process is usually carried out using the following conditions: temperature 400–450 °C; pressure 1–2 atmospheres; using a catalyst, vanadium pentoxide. Explain why these conditions are chosen.

_____

_____

_____

_____ [4 marks]

2.

Higher Tier only

Substance **X** is in equilibrium with substance **Y**.

X(g) ⇌ 2**Y**(g)

Temperature in °C	Conversion to Y at different pressures		
	1 atm	5 atm	10 atm
20	15%	10%	2%
50	35%	24%	4%
100	60%	41%	7%

Use the data in the table to:

a Identify whether the forward reaction is exothermic or endothermic. Explain your answer.

_____

_____

_____ [3 marks]

b Explain the variation in percentage conversion with pressure.

_____

_____

_____ [3 marks]

3. Explain what is meant by the term 'equilibrium'. Use the Haber process as an example.

Higher Tier only

Synoptic

_____

_____

_____

_____

_____

_____

_____ [6 marks]

Synoptic

In the exam, some of the marks will be for connecting your knowledge from different areas of chemistry. In this question you will need to apply your knowledge of equilibrium to explain the Haber process (Section 10).

Crude oil and hydrocarbons

1. Crude oil is a mixture of hydrocarbons.

a Name the two elements found in every hydrocarbon.

_____ [2 marks]

b Crude oil is a fossil fuel. Explain what is meant by the term 'fossil fuel'.

_____

_____ [2 marks]

c Explain what the original source of energy in fossil fuels is.

_____

_____ [2 marks]

d Suggest a reason why fossil fuels are called 'non-renewable fuels'.

_____ [1 mark]

Structure and formulae of alkanes

1. **a** Which molecule is an alkane? Tick **one** box.

☐ C_2H_4 ☐ C_3H_6 ☐ C_4H_{10} ☐ C_5H_8 [1 mark]

b Describe the main characteristics of the alkane molecules.

_____

_____ [2 marks]

2. Look at the first four alkane molecules:

$$H-C-H$$ with H above and H below (methane)

$$H-C-C-H$$ with H H above and H H below (ethane)

$$H-C-C-C-H$$ with H H H above and H H H below (propane)

$$H-C-C-C-C-H$$ with H H H H above and H H H H below (butane)

a Circle the ethane molecule.   [1 mark]

Remember
Look for patterns in the molecules shown.

b The next molecule in the series has 5 carbon atoms. Explain why its formula is C_5H_{12}.

_____

_____ [2 marks]

c Predict the formula of the alkane containing 7 carbon atoms.

_____ [1 mark]

3. The diagram shows some hydrocarbon molecules.

A

$$H-C-C-C-C-C-H$$ with H H H H H above and H H H H H below

B

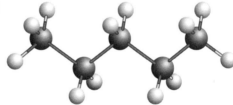

C

H–C–H (top)
$$H-C-C-C-C-H$$ with H H H above and H H H H below

D

H–C–H (top)
$$H-C-C-C-C-C-H$$ with H H H H above and H H H H H below

Which molecule is the odd one out? Explain your answer.

_____

_____ [2 marks]

Fractional distillation and petrochemicals

1. Crude oil is a mixture of hydrocarbons which can be separated into different fractions by fractional distillation.

   a Describe how fractional distillation separates crude oil into different fractions.

_____

_____

_____

_____

_____ [5 marks]

   b Explain why bitumen has a higher boiling point than kerosene.

_____

_____

_____

_____ [3 marks]

2. The table shows the percentages of the fractions in crude oil from different oil wells.

Fraction	% of fractions in oil well A	% of fractions in oil well B
LPC	6	2
Petrol	10	11
Paraffin	15	20
Diesel	7	5
Heating oils	6	4
Fuel oil	14	12
Bitumen	42	46

Maths A barrel of crude oil from well **A** has a mass of 150 kg, whereas a barrel from well **B** has a mass of 142 kg. Calculate which well produces the most paraffin.

_____

_____ [3 marks]

Properties of hydrocarbons

1. The properties of hydrocarbon fractions are listed in this table.

Number of carbon atoms	Boiling point in °C	Viscosity Relative scale 1 (low) to 10 (high)	Flammability Relative scale 1 (poor) to 10 (good)
3–4	Below 30	Gas	10
6–10	80–120	2	8
9–16	110–190	3	6
12–19	130–230	6	5
20–27	240–340	8	3
28–32			

Use the data from the table to answer the questions.

Maths **a** Complete the table for 28–32 carbon atoms.

[3 marks]

> **Maths**
> Look for trends in the data shown to predict the missing values.

b Describe the trend in flammability as the number of carbon atoms increases.

_____ [1 mark]

c Describe the trend seen in boiling point as the number of carbon atoms increases.

_____ [1 mark]

d Explain the trend in viscosity as the number of carbon atoms increases.

_____

_____

_____

Analysing questions

Use your knowledge of forces between molecules to answer this question.

_____ [3 marks]

Combustion of fuels

1. **a** The hydrocarbon C_6H_{14} was burned in air and complete combustion occurred. Which equation represents the reaction correctly? Tick **one** box.

☐ $C_6H_{14} + O_2 \rightarrow 6CO + 7H_2$

☐ $C_6H_{14} + 13O \rightarrow 6CO + 7H_2O$

☐ $C_6H_{14} + 9.5O_2 \rightarrow 6CO_2 + 7H_2O$

☐ $C_6H_{14} + 3O_2 \rightarrow 6CO + H_2$ [1 mark]

b Write a balanced equation for the complete combustion of propane, C_3H_8.

_____ [2 marks]

c Sam makes the following observations when using a Bunsen burner to heat some water.

Remember

Methane is the gas burned in a Bunsen burner.

• When the air hole is open the flame is blue.

• When the air hole is closed the flame is yellow.

Explain these observations.

_____

_____

_____

_____ [4 marks]

2. A fuel is a substance that is burned to produce heat. Describe and explain the factors you would consider when choosing a suitable fuel for cooking when camping.

Synoptic

Analysing questions

When camping you will be using a small portable stove to cook, so you need to take this into account in your answer, as well as consider how the fuel actually burns.

_____

_____

_____

_____ [4 marks]

Cracking and the alkenes

1. The laboratory apparatus used for cracking $C_{18}H_{38}$ is shown below.

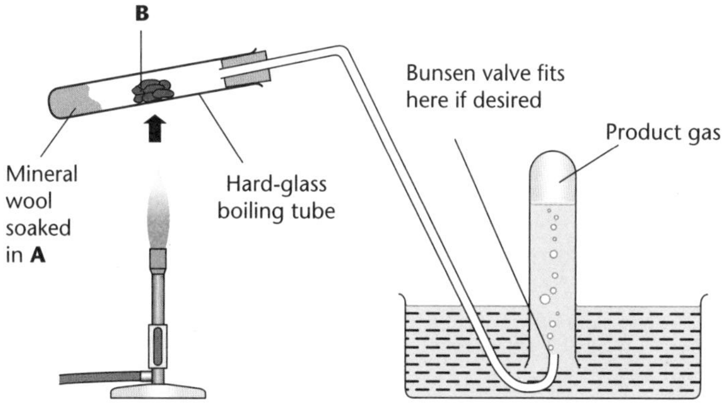

Required practical

B

Bunsen valve fits here if desired

Product gas

Mineral wool soaked in **A**

Hard-glass boiling tube

a Identify the names of substances **A** and **B**.

A is _____ **B** is _____ [2 marks]

b Describe what is meant by the term 'cracking'.

_____ [1 mark]

c The equation for the following reaction is:

$$C_{18}H_{38} \rightarrow C_6H_{14} + C_4H_8 + \underline{\quad}C_3H_6 + C_2H_4$$

Complete the equation by balancing it [1 mark]

and identify which of the products are alkenes.

_____ [1 mark]

Remember

To balance an equation, count the number of atoms in each molecule on both sides of the arrow. Remember, you can only add numbers in front of the molecules. You cannot change numbers within the molecules.

d Describe a chemical test that will show if the cracking reaction has been successful.

_____

_____

_____ [3 marks]

2. Describe what happens during steam cracking. Explain why it is an important industrial process.

_____

_____

_____

_____ [4 marks]

Structure and formulae of alkenes

1. **a** Which of these molecules is an alkene? Tick **one** box. [1 mark]

☐ C_2H_6 ☐ C_3H_6 ☐ C_4H_{10} ☐ C_5H_8

b Explain why the methene molecule does not exist.

_____

_____

_____ [3 marks]

2. Look at the first four alkene molecules.

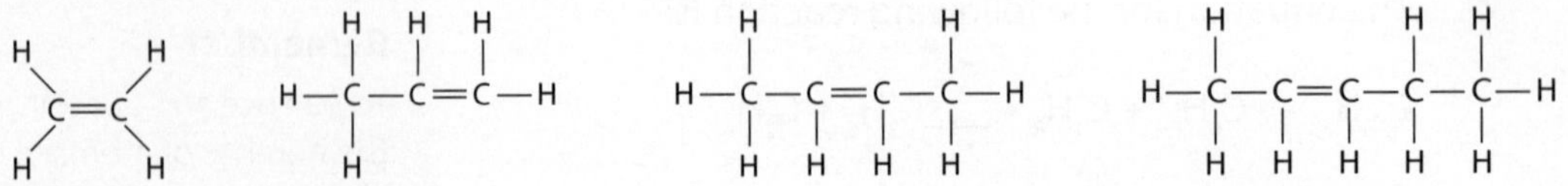

$$\begin{array}{c} H \\ \diagdown \\ C = C \\ \diagup \quad \diagdown \\ H \qquad H \end{array} \qquad H-\underset{\underset{H}{|}}{C}-C=C-H \qquad H-\underset{\underset{H}{|}}{\overset{H}{C}}-C=C-\underset{\underset{H}{|}}{\overset{H}{C}}-H \qquad H-\underset{\underset{H}{|}}{\overset{H}{C}}-C=C-\underset{\underset{H}{|}}{\overset{H}{C}}-\underset{\underset{H}{|}}{\overset{H}{C}}-H$$

a Circle the butene molecule. [1 mark]

b The next molecule in the series has 6 carbon atoms. Explain why the formula is C_6H_{12}.

_____

_____

_____ [2 marks]

c Predict the formula of an alkene containing 10 carbon atoms.

_____ [1 mark]

3. Look at this margarine tub.

Explain the meaning of the terms 'unsaturated' and 'saturated' in this context.

This margarine is better for you as it's high in unsaturated fats and low in saturated fat.

_____

_____

_____ [2 marks]

Reactions of alkenes

1. **a** When propene reacts with iodine, an addition reaction takes place. Which structure shows the product of this reaction? Tick **one** box.

Ⓐ
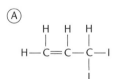

Ⓑ
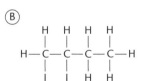

Ⓒ
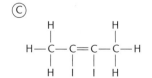

Ⓓ
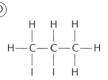

☐ A ☐ B ☐ C ☐ D [1 mark]

b When propene reacts with steam, an alcohol is formed. Which structures show the possible products? Tick **two** boxes. [2 marks]

Ⓐ
```
    H   H   H
    |   |   |
H — C — C — C — H
    |   |   |
    H  OH   H
```

Ⓑ
```
    H   H   H
    |   |   |
    C = C — C — H
    |   |   |
    H  OH   H
```

Ⓒ
```
    H   H   H
    |   |   |
H — C — C — C — H
    |   |   |
    OH  H   H
```

Ⓓ
```
    H   H   H
    |   |   |
H — C = C — C — H
    |   |   |
    OH  H   H
```

☐ A   ☐ B   ☐ C   ☐ D   [2 marks]

---

2. Alkenes react with oxygen but they tend to burn with smoky flames.

a Complete the equation showing the incomplete combustion of ethene.

$C_2H_4 + O_2 \rightarrow$ _____ + _____ H_2O [2 marks]

b Suggest a reason why the flame is smoky.

_____

[1 mark]

---

3. Compare the structures of alkanes and alkenes. Explain how their structure affects their chemical reactivity.

_____

_____

_____

_____

_____

_____

_____

_____

[6 marks]

> **Remember**
>
> When comparing the structures of alkanes and alkenes, draw out displayed formula of specific examples so you can compare the bonds and the numbers of carbon atoms and hydrogen atoms. For chemical reactivity, include combustion and other chemical reactions, for example with halogens.

Structure and formulae of alcohols

1. Propanol is an alcohol. Which of these structures shows propanol? Tick **one** box. [1 mark]

Ⓐ
H   H   H
| | |
H—C—C—C—OH
| | |
H   H   H

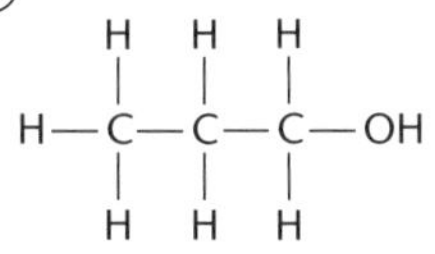

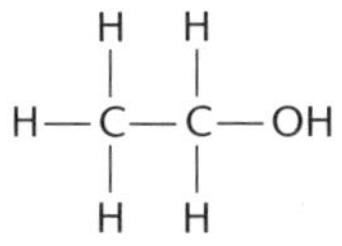

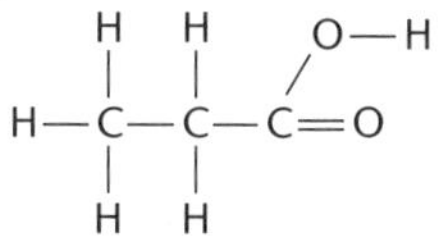

Ⓑ
H   H
| |
H—C—C—OH
| |
H   H

Ⓒ
H   H   O—H
| | /
H—C—C—C=O
| |
H   H

Ⓓ
        OH
        /
H—C=O

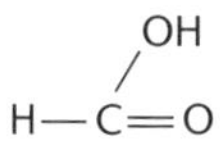

☐ A   ☐ B   ☐ C   ☐ D   [1 mark]

2. Ethanol can be made by fermentation. It can also be made, on an industrial scale, by the reaction between ethene and steam.

a Complete the word equation for the fermentation reaction.

Glucose → _____ + _____   [1 mark]

b Write the symbol equation with state symbols for the industrial reaction.

_____   [2 marks]

c Explain why the reaction conditions including temperature and pH need monitoring carefully during a fermentation reaction.

_____

_____

_____

_____

_____   [4 marks]

Uses of alcohols

1. Organic compounds have many uses.

a Which of the following uses methanol as a feedstock? Tick **two** boxes.

☐ Food flavouring ☐ Vinegar

☐ Alcoholic drinks ☐ Adhesives [2 marks]

b Ethanol can be used as a fuel. Write the balanced symbol equation for the combustion of ethanol.

_____ [2 marks]

Required practical

c Esters are sweet-smelling organic compounds that are used as perfumes and food flavouring. Describe how you would make an ester. Use an example in your answer.

_____

_____

_____ [4 marks]

2. Substance **A** is a hydrocarbon containing 10 carbon atoms. When heated with a catalyst it forms **B**, which has 2 carbon atoms, and one other compound **C**. **B** reacts to form **D**. When sodium is added to **D**, the gas produced pops when a lighted splint is inserted.

Synoptic

Identify substances **A**, **B**, **C** and **D**. Give details of the reactions that took place. Include equations where appropriate.

Literacy

In this question you will be assessed on using good English, organising information clearly and using specialist terms where appropriate. Use your knowledge and understanding of organic chemistry to answer it.

_____

_____

_____

_____

_____ [6 marks]

Carboxylic acids

1. Butanoic acid is a carboxylic acid.

a Which of these structures is that of butanoic acid? Tick **one** box.

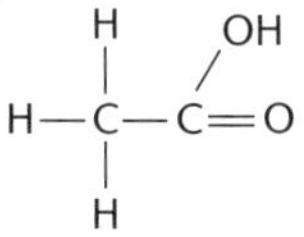

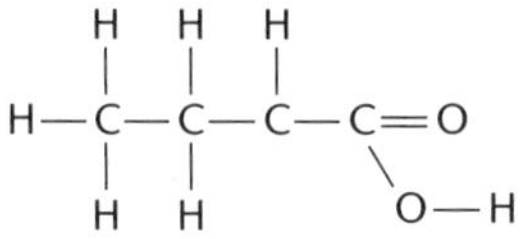

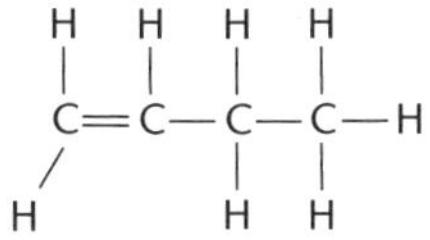

☐ A ☐ B

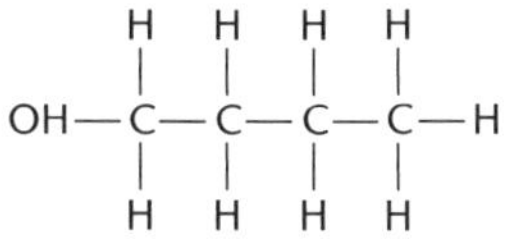

☐ C ☐ D [1 mark]

b Butanoic acid is formed by the oxidation of which organic compound? Circle the name of **one** compound.

butane   butanol   (poly)butene   butene   [1 mark]

2. **a** Describe what happens when methanoic acid is mixed with sodium carbonate.

_____ [1 mark]

Higher Tier only **b** Explain why methanoic acid is a weak acid, but nitric acid is a strong acid.

_____

_____

_____

_____ [4 marks]

3. Compare the structure, solubility and chemical properties of the carboxylic acids with those of the alkanes.

Use these words in your answer:

ethanoic acid   ethane   functional group

solubility   combustion   esterification

_____

_____

_____

_____

_____

_____   [6 marks]

Addition polymerisation

1. **a** Which of the following statements about poly(ethene) is true? Tick **one** box.

☐ It is formed by the addition polymerisation of ethane.

☐ It is made by condensation polymerisation.

☐ Ethene monomers are used to make poly(ethene).

☐ Hydrogen gas is given off during the production of poly(ethene).   [1 mark]

b Propene is used in the manufacture of the plastic (poly)propene. Draw the bonds to complete the displayed formulae of propene and poly(propene).

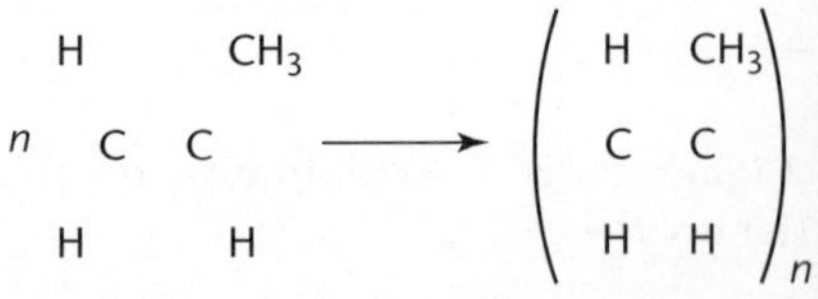

$$n \quad \begin{matrix} H & CH_3 \\ C & C \\ H & H \end{matrix} \longrightarrow \left(\begin{matrix} H & CH_3 \\ C & C \\ H & H \end{matrix} \right)_n$$

[2 marks]

c Describe what happens to the propene molecules during polymerisation.

_____

_____   [2 marks]

2. The diagram below shows part of a molecule from polymer A and part of one from polymer B. Both polymers were made by addition polymerisation.

Polymer A   Polymer B

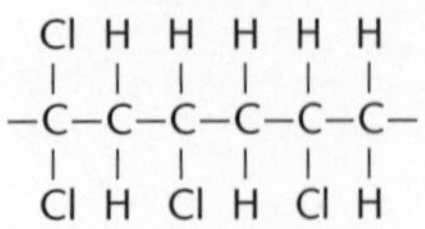

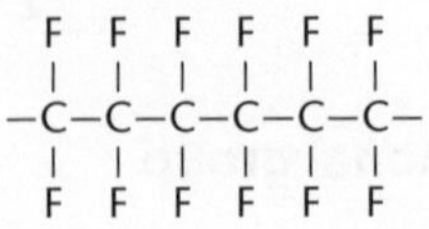

$$\begin{matrix} Cl & H & H & H & H & H \\ | & | & | & | & | & | \\ -C & -C & -C & -C & -C & -C- \\ | & | & | & | & | & | \\ Cl & H & Cl & H & Cl & H \end{matrix} \qquad \begin{matrix} F & F & F & F & F & F \\ | & | & | & | & | & | \\ -C & -C & -C & -C & -C & -C- \\ | & | & | & | & | & | \\ F & F & F & F & F & F \end{matrix}$$

Compare the two polymers and explain how they were made.   [4 marks]

Polymer A has the following pattern: the first C atom has 1 H and 1 Cl attached; the second C atom has 2 Hs attached. Polymer B has 2 F atoms attached to every C atom.

Polymerisation happens when small molecules, called monomers, are joined together to form large molecules.

Polymer A is made from the monomer $_2HC=CClH$

Polymer B is made from the monomer $_2FC=CF_2$

During polymerisation the C=C double bonds open up and are replaced by C–C single bonds which join the monomers together.

Marks gained: [4 marks]

Condensation polymerisation

1. Nylon is made by condensation polymerisation.

Higher Tier only **a** Explain what is meant by 'condensation' polymerisation.

_____

_____ [2 marks]

b This diagram shows the structure of a nylon polymer.

Circle the repeating unit. [1 mark]

c The diagram below shows the two monomers used to make a polymer that is similar to nylon.

Draw the structure of this polymer. Show one repeating unit and use the letter '*n*' to indicate that it is repeated many times. [3 marks]

85

2.

Higher Tier only

Polymers are produced by two different processes. The equation for the reaction to produce poly(ethene) is:

$$n\, \overset{\displaystyle H \quad H}{\underset{\displaystyle H \quad H}{C=C}} \longrightarrow \left(\overset{\displaystyle H \quad H}{\underset{\displaystyle H \quad H}{C-C}} \right)_n$$

The equation for the reaction to produce poly(ester) is:

$$n\text{HO}-\square-\text{OH} + n\text{HOOC}-\square-\text{COOH} \longrightarrow \left(\square-\text{OOC}-\square-\text{COO}\right)_n + 2n\text{H}_2\text{O}$$

Compare these two different polymerisation processes.

_____

_____

_____ [4 marks]

> **Command words**
>
> The command word 'compare' means to describe differences *and* similarities between the two processes.

Amino acids

1.

Higher Tier only

The displayed formula for alanine is shown below.

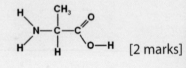

a Circle the **two** functional groups alanine has. [2 marks]

b Name the molecule that is eliminated during polymerisation.

_____ [1 mark]

c Write down the formula of the repeating unit.

_____ [1 mark]

2.

Higher Tier only

Describe how proteins are formed from amino acids. Use glycine, H_2NCH_2COOH, as an example.

_____

_____

_____ [5 marks]

DNA and other naturally occurring polymers

1. Draw a circle round the monomer that produces the natural polymer cellulose.

amino acid **glucose** **monosaccharide** **starch** [1 mark]

2. This diagram shows a simplified structure for glucose:

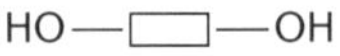

HO —☐— OH

Glucose is a monosaccharide.

The diagram below shows a section of a polysaccharide:

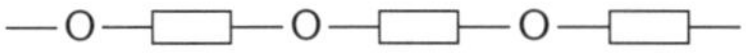

—O—☐—O—☐—O—☐—

Describe how monosaccharide molecules join together to form polysaccharides such as starch.

_____

_____

_____

_____ [4 marks]

Analysing questions

A good way to approach this question is to draw the monomers in a line and then put a circle round the atoms that are going to be eliminated.

3. **a** Explain why DNA is an important molecule.

_____ [1 mark]

b Describe the structure of a molecule of DNA.

_____

_____ [2 marks]

c The diagram below shows four nucleotides, each including a different base group.

| sugar | acid | phosphate |
| base | sulfate | sucrose |

Label the diagram using words from the box. [3 marks]

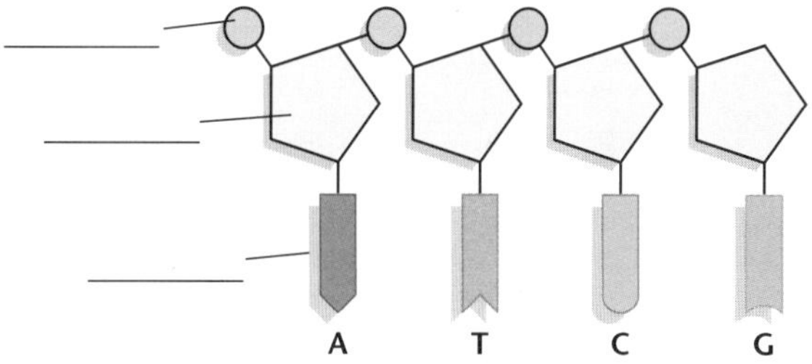

A T C G

d Circle the nucleotide with a thymine base. [1 mark]

Pure substances, mixtures and formulations

1. Mohammed suspects that his supply of distilled water has been contaminated, so he decides to measure its boiling point.

a Why does Mohammed measure the boiling point?

_____

_____ [2 marks]

b His test result is 102 °C. What does this result tell Mohammed?

_____ [1 mark]

c The contaminant is separated from the water and is found to consist of white crystals. Suggest how this has been done.

_____ [1 mark]

d It is found that the contaminant has a melting point of 175–186 °C. What does this mean about the purity of the contaminant?

_____ [2 marks]

e Mohammed wants to identify the contaminant. What should he do next?

_____ [1 mark]

2. **a** Paracetamol is a formulation. Explain why the substances in formulations have to be mixed in the correct proportions.

_____

_____ [2 marks]

b When formulations like paracetamol are made on an industrial scale, quality assurance tests are carried out. Explain why.

_____ [1 mark]

c The table shows the quality assurance test results for some samples of paracetamol tablets.

Table 1

Sample	A	B	C	D
% paracetamol	71.4	71.5	71.4	76.6
% binder	2.0	2.0	2.0	2.0
% corn starch	5.0	5.0	5.0	1.0
% other ingredients	22.6	22.5	22.6	20.4

Which sample would get rejected and why?

_____

_____ [2 marks]

Chromatography

1. Chromatography is used to separate a mixture of amino acids using ethanol as a solvent. The diagram shows the result.

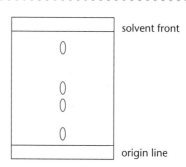

a How many different amino acids were in the mixture?

_____ [1 mark]

b On the diagram, write a letter **S** next to the amino acid that is most soluble in ethanol. [1 mark]

c Explain why the solvent used in this experiment was ethanol and not water.

_____ [1 mark]

2. Ershad is going to use chromatography to identify the components in a mixture of coloured dyes. The diagram shows the apparatus.

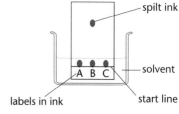

a Explain why Ershad needs to start again with a fresh piece of chromatography paper.

_____

_____

Command words

'Explain why' means to make the reasons why something is happening clear.

_____

_____ [4 marks]

The diagram shows the results of Ershad's experiment.

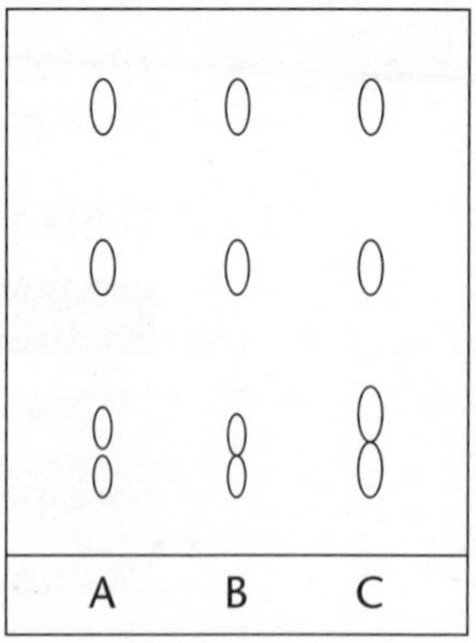

b Analyse the chromatogram. What does it tell us about the mixture?

_____ [1 mark]

c How could Ershad identify the actual dyes in the mixture.

_____

_____ [2 marks]

Chromatography and R_f values

1. The substances on a chromatogram can be identified by calculating R_f values.

Maths

a Define the meaning of 'R_f value'.

Required practical

_____

_____ [2 marks]

b Write the expression used to calculate an R_f value.

_____ [1 mark]

Remember
You need to learn the expression for R_f – it may not be given to you in the exam.

2. The R_f values of four substances are listed in the table.

Substance	R_f value	Distance travelled in mm
A	0.36	18
B	0.92	46
C	0.11	
D	0.56	

Maths
When calculating R_f values, your answer will always be less than 1.

a Calculate how far the solvent travelled.

_____

_____ [2 marks]

Maths
You will need to rearrange the R_f expression.

b Complete the table by calculating the distances travelled by substances **C** and **D**.

Give your answers to 2 significant figures. [2 marks]

Maths
You will need to use your answer from part (a) and rearrange the R_f expression.

3. The diagram shows a typical chromatogram.

Maths

Required practical

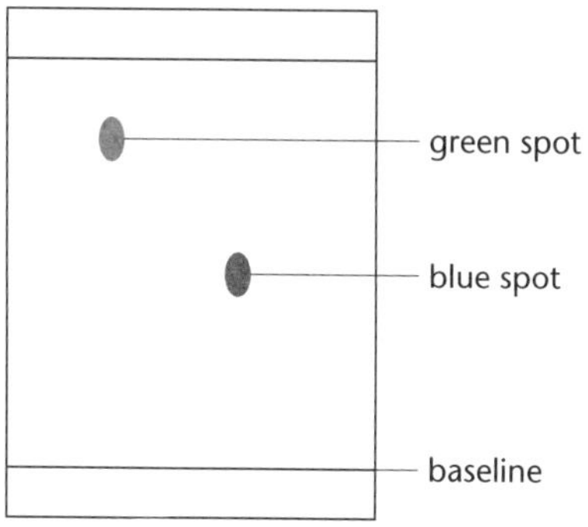

a Draw lines on the chromatogram to show how far:

- the solvent travelled. Label it 'S'. [1 mark]

- the green spot travelled. [1 mark]

green spot

blue spot

baseline

b Calculate the R_f value of the green dye.

_____ [2 marks]

c Why do R_f values have no units?

_____ [1 mark]

Tests for common gases

1. The following observations were made when testing an unknown gas, **X**.

- Pale green.
- Turns damp blue litmus paper red and bleaches it.

- Distinct smell.
- Extinguishes a glowing splint.

a Give one piece of evidence that shows gas **X** to be acidic.

_____ [1 mark]

b Name gas **X**. _____ [1 mark]

2. Some students are investigating the reactions of an unknown solid, **Y**. When they mix **Y** with acid they notice that a gas, **Z**, is given off. They test the gas. The results are shown in the table below.

Test	Result
Hold a lighted splint to the open end of the tube.	It remains alight.
Put a glowing splint into the tube.	It goes out.
Put a drop of limewater on a glass rod and put it in the test tube.	It goes milky.
Hold damp blue litmus paper to the open end of the test tube.	It goes red.

a Suggest how these results will help Eric to identify **Y**.

_____

_____

_____

_____ [4 marks]

Literacy

This question is worth 4 marks, so it is important to comment on the results of all the tests rather than just stating the answer.

Required practical

b The students want to check their results. This time they will collect four test tubes full of the gas. They use the following pieces of equipment:

- Conical flask
- Delivery tube
- Water trough
- Test tube
- Bung

Draw a labelled diagram in the box to show how they will set up the equipment.

[3 marks]

Flame tests

1. **a** An ionic chloride produces a lilac colour in a flame. Identify the correct compound. Tick **one** box.

  ☐ Calcium chloride   ☐ Lithium chloride

  ☐ Sodium sodium   ☐ Potassium chloride   [1 mark]

b Circle the metal ion that gives a crimson flame colour.

lithium   sodium   potassium   calcium   [1 mark]

2. Keira is trying to identify three unknown samples using flame tests. Each time she does the test she gets the same result – a yellow flame. She notices that other groups are getting different results. Suggest **one** reason why Keira's results are all the same.

_____

_____   [2 marks]

3. Dilute hydrochloric acid was added to a green solid compound, **X**, and fizzing was observed. Compound **X** gave a green flame when a flame test was carried out.

Required practical

a Identify compound **X**.

_____   [1 mark]

b Complete the equation for the reaction of **X** with an acid and add the state symbols.

$CuCO_3$(___) + ___HCl(___) → $CuCl_2$(___) + H_2O(___) + CO_2(___)   [3 marks]

Metal hydroxides

1. The table shows the results observed by Jack and Joan having done some flame tests.

Required practical

Sample	Flame test	Test with NaOH	Test for CO_2	Cation present
A	Yellow	No change	No change	
B	Orange-red	White ppt	Milky	
C	Green	Blue ppt	No change	

a Complete the table.   [3 marks]

b Identify sample **B**.   _____   [1 mark]

c Explain why it is not possible to identify compounds **A** and **C** using these results.

_____

_____   [2 marks]

2.

Required practical

An ionic compound formed a white precipitate with sodium hydroxide. The precipitate did not dissolve in excess sodium hydroxide. Describe how you could identify the metal ion present.

_____

_____ [2 marks]

Practical

You should be able to identify the ions present in unknown single ionic compounds given a description of the chemical test(s) used.

3.

a Describe what happens when sodium hydroxide is added to copper sulfate solution.

_____ [1 mark]

b Write a balanced equation for the reaction.

_____ [3 marks]

4. Aqueous sodium hydroxide was added to a metal chloride solution and a light green precipitate was formed.

Required practical

a Identify the metal ion present. _____ [1 mark]

Higher Tier only

b Complete this ionic equation for the reaction.

$Fe^{2+} + \_\_OH^- \rightarrow _____$ [2 marks]

Tests for anions

1. Very dilute solutions of iron(III) chloride and iron(III) sulfate are both pale brown. Explain how you would distinguish between the two solutions.

Required practical

_____

_____

_____ [4 marks]

2.

Required practical

Khalid was given three unlabelled bottles containing crystals of sodium chloride, sodium sulfate and calcium chloride, respectively, and was asked to label them. Describe a method to identify each solid positively using the lowest possible number of reactions. Give the expected observations in each case when a reaction takes place.

Literacy

You need to use your knowledge and understanding of chemical tests to answer this question. For 6-mark questions like this you will be marked on your use of English, so remember to write in complete sentences.

_____

_____

_____

_____

_____

_____

_____

_____ [6 marks]

Instrumental methods

1. The diagram shows a gas chromatogram.

a Which substance has the longest retention time?

_____ [1 mark]

b Which substances are present in equal amounts?

_____ [2 marks]

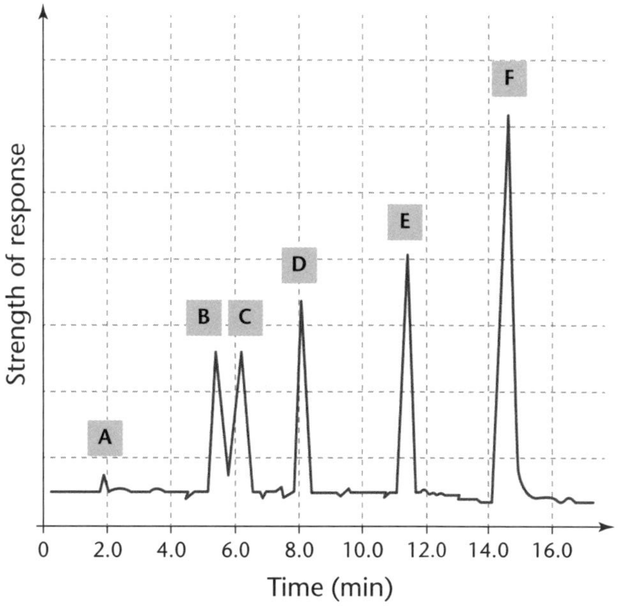

c Explain why an instrumental method of carrying out chromatography may have advantages over paper chromatography.

_____

_____

_____ [3 marks]

2. Electrophoresis is used to separate fragments of DNA. The diagram shows the results of an electrophoresis experiment.

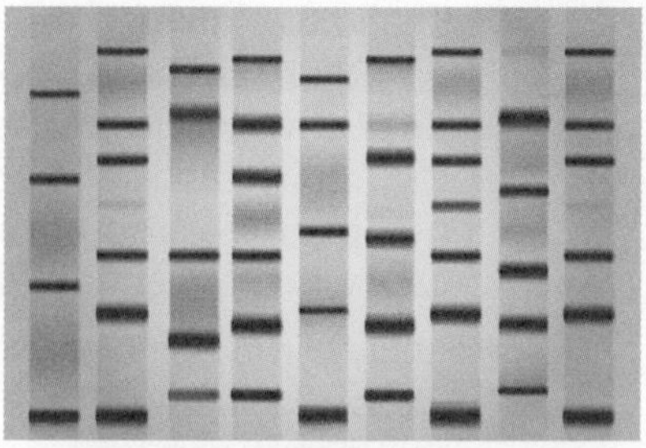

Suggest why this type of DNA analysis is useful in the fight against crime.

_____

_____ [2 marks]

Command words

If the command word 'suggest' is used, you need to apply your knowledge and understanding of the topic to a new situation.

3. In simple terms, the peaks of a mass spectrum represent the molecular mass of a fragment of a molecule if it is either whole or split. The diagram shows part of the mass spectrum of the products of a cracking reaction.

Synoptic

a Look at the diagram and suggest which alkenes are present. The A_r of C is 12 and the A_r of H is 1.

_____ [2 marks]

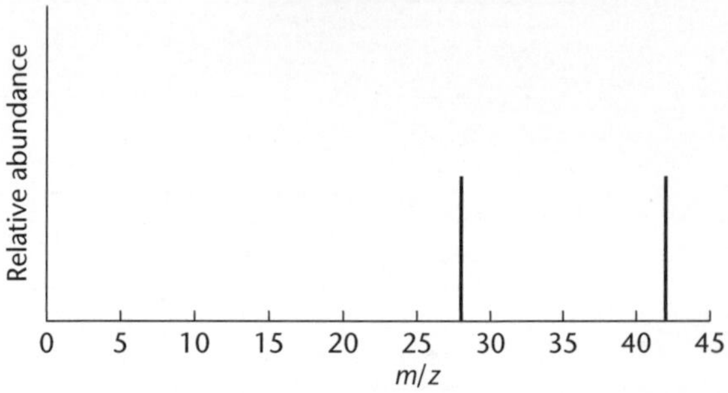

b What is the chemical test for an alkene?

_____ [1 mark]

c Compare the results of the chemical test with those from the mass spectrum.

_____

_____

_____ [3 marks]

Synoptic

In the exam, some of the marks will be for connecting your knowledge from different areas of chemistry. Here you need to link your understanding of instrumental methods with your knowledge of organic chemistry.

Flame emission spectroscopy

1. **a** Describe how you would obtain a flame emission spectrum.

_____

_____ [2 marks]

2. Give one advantage of flame emission spectroscopy over flame tests.

_____ [1 mark]

3. Flame emission spectroscopy can be used to analyse metal ions in solution. The diagram shows the flame emission spectra of the Group 1 metal ions and an unknown metal ion, **X**.

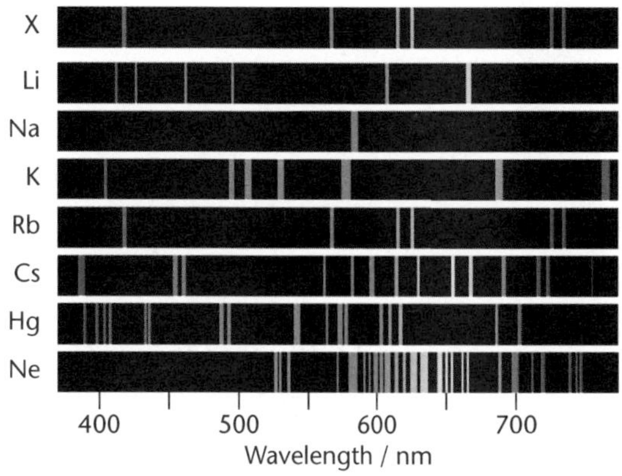

a Identify the ions present in the sample of **X**. Give a reason for your answer.

_____

_____ [2 marks]

b A technician suspects that another solution, **Y**, which should contain sodium chloride, has been contaminated with potassium ions. Describe what the technician should do to see if they are right.

Command words

'Identify' means to name or otherwise characterise. The answer is usually short.

_____

_____

_____

_____ [3 marks]

c Explain why it might not be a good idea for the technician to carry out a flame test.

_____

_____ [2 marks]

The Earth's atmosphere – now and in the past

1. The bar chart shows the proportions of gases in the atmosphere today.

a Name the gas shown in bar **B**.

_____ [1 mark]

b Bar **C** represents a mixture of different gases. Name **one** gas in this mixture.

_____ [1 mark]

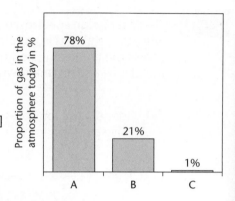

2. The table shows the composition of the atmospheres of Mars, Venus and Earth.

Maths

Gas	Venus atmosphere %	Mars atmosphere %	Earth
Carbon dioxide	96.5	95.0	0.039
Nitrogen	3.5	3.0	78.1
Oxygen	Traces	None	21.0
Other	None	2.0	0.861

a Calculate the ratio of the amount of carbon dioxide gas to all other gases plus nitrogen found on Mars.

_____

_____ [2 marks]

b Calculate how many times more nitrogen there is on Earth compared with Mars.

_____

_____ [3 marks]

c Calculate how many times more carbon dioxide there is on Venus compared with Earth. Give your answer to 3 significant figures.

_____

_____ [2 marks]

Maths

To calculate a ratio, first write the percentage as a fraction, then divide the fractions by the smallest number and write down the ratio. Remember to show your working; you may get some marks even if your final answer is wrong.

3. There are several theories about how the Earth's early atmosphere and oceans were formed. One theory involves intense volcano activity, which released gases during the first billion years of Earth's existence. Water vapour condensed to produce the oceans. Evaluate this theory.

_____

_____

_____

_____

_____

_____ [6 marks]

Changes in oxygen and carbon dioxide

1. The diagram shows how scientists think that the levels of oxygen have built up in the atmosphere over billions of years.

The unit on the x axis (Ga) means 'billions of years'.

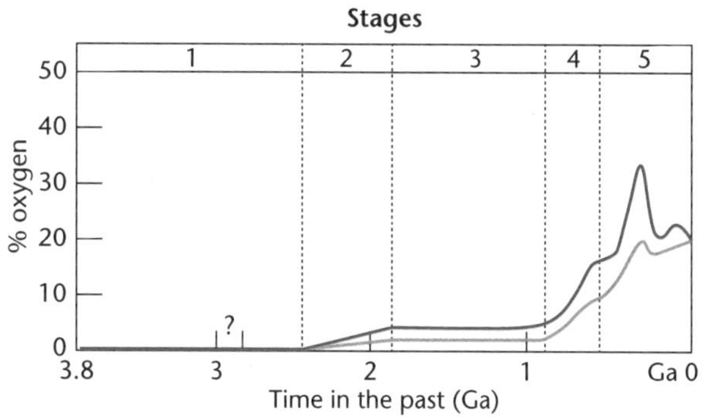

a The two lines drawn on the graph show the range of estimates. Suggest a reason why this graph uses estimated data.

_____ [1 mark]

b Suggest sources of evidence that scientists have used to base their estimaties on.

_____

_____ [2 marks]

c Algae and plants produced the oxygen that is in our atmosphere today as a by-product of photosynthesis. Complete the equation for photosynthesis.

$\_\_CO_2(g) + \_\_H_2O(g) \rightarrow C_6H_{12}O_6(aq) + \_\_O_2(g)$ [2 marks]

d Use the graph at the start of this question to identify when the first oxygen-producing organisms started to evolve.

_____ [1 mark]

e Describe what scientists think happened to the oxygen level in the atmosphere over the last two billion years.

_____

_____ [3 marks]

2. The composition of the Earth's atmosphere today is different from its early atmosphere. Describe how the proportions of gases have changed. Suggest why scientists might be uncertain about this information.

Literacy
In this question you will be assessed on using good English, organising information clearly and using specialist terms where appropriate. Use your knowledge and understanding of gases in the atmosphere to answer it.

_____

_____

_____

_____

_____ [6 marks]

Greenhouse gases

1. Greenhouse gases in the atmosphere maintain temperatures on Earth high enough to support life.

a Circle the gas that is **not** a greenhouse gas.

water vapour   **nitrogen**   **methane**   **carbon dioxide**   [1 mark]

b Describe what is meant by the 'greenhouse effect'.

Literacy
You will need to include ideas about long and short wavelengths of radiation to gain full marks in this question.

_____

_____

_____

_____ [4 marks]

2. The diagram shows the carbon dioxide levels in the atmosphere over the last 400 000 years.

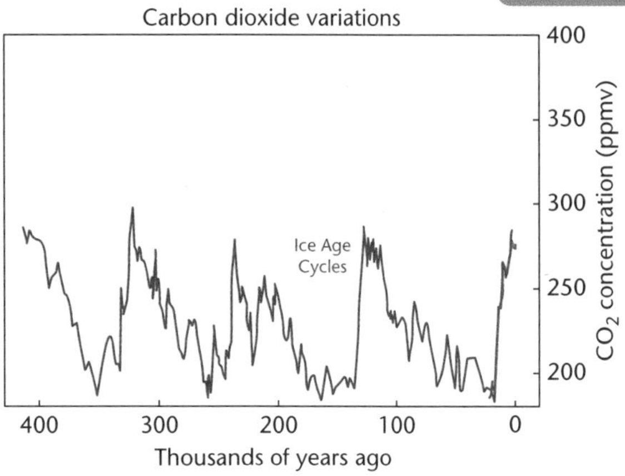

Maths **a** Describe the trends in the levels of carbon dioxide gas during the last 400 000 years.

_____ [1 mark]

This second diagram shows changes in carbon dioxide levels in the atmosphere over the last 1000 years.

Carbon dioxide variations

Maths **b** Describe what has happened to the carbon dioxide levels over the last 1000 years.

_____

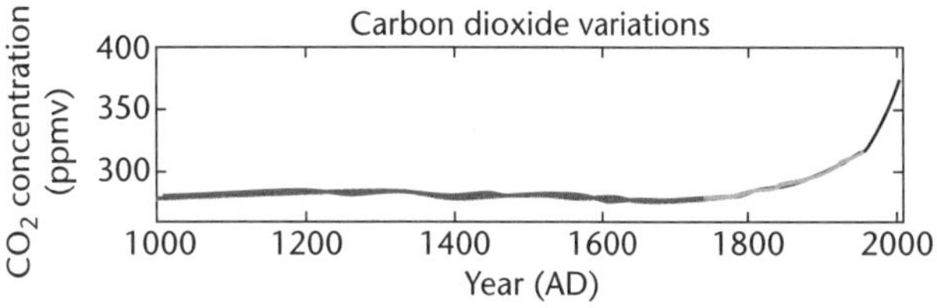

_____ [2 marks]

c Calculate the percentage *increase* in the concentration of CO_2 between 1800 and 2000. Give your answer to 3 significant figures.

_____

_____ [3 marks]

Maths

Use the graph or your answer to part (b). First calculate the actual increase and then calculate it as a percentage.

d Suggest **two** reasons for this increase in CO_2 levels.

_____

_____ [2 marks]

e Give **one** effect of increased levels of CO_2.

_____ [1 mark]

Global climate change

1. **a** Describe what is meant by the term 'global warming'.

_____ [1 mark]

b An increase in global temperatures will cause climate change. What is one possible effect of change? Tick **one** box.

☐ Sea levels rise. ☐ More diseases.

☐ Longer winters. ☐ New power stations. [1 mark]

2. **a** Scientists collect, analyse and interpret climate data. Explain why it is important for the resulting evidence to be peer-reviewed.

> **Remember**
>
> 'Peer review' is the process in which a scientist's work is scrutinised by other scientists working in the same area of research.

_____

_____ [2 marks]

b Suggest **one** reason why is it difficult for scientists to model global warming.

_____ [1 mark]

3. If global warming continues there will be big changes in the environment. Describe **two** environmental changes that could take place and discuss their possible impact.

_____

_____

_____ [4 marks]

Carbon footprint and its reduction

1. Which definition of a 'carbon footprint' is correct? Tick **one** box.

☐ The total amount of carbon dioxide emitted during the making of a product.

☐ The total amount of carbon dioxide and other greenhouse gases emitted during the making of a product.

☐ The total amount of carbon dioxide and other greenhouse gases emitted over the full life cycle of a product, service or event.

☐ The total amount of carbon dioxide emitted over the full life cycle of a product, service or event. [1 mark]

2. Explain how individual households can reduce their carbon footprint.

_____

_____

_____

_____ [4 marks]

Analysing questions

You have been asked to 'explain' methods by which households can reduce their carbon footprint. Don't just _list_ measures that could be taken, _explain_ how the measure will work.

3. The government is to trying to reduce the nation's carbon footprint. Suggest some advice you would give the government.

_____

_____

_____ [3 marks]

4. The diagram shows the global population and the concentration of methane gas in the atmosphere between 1984 and 2004.

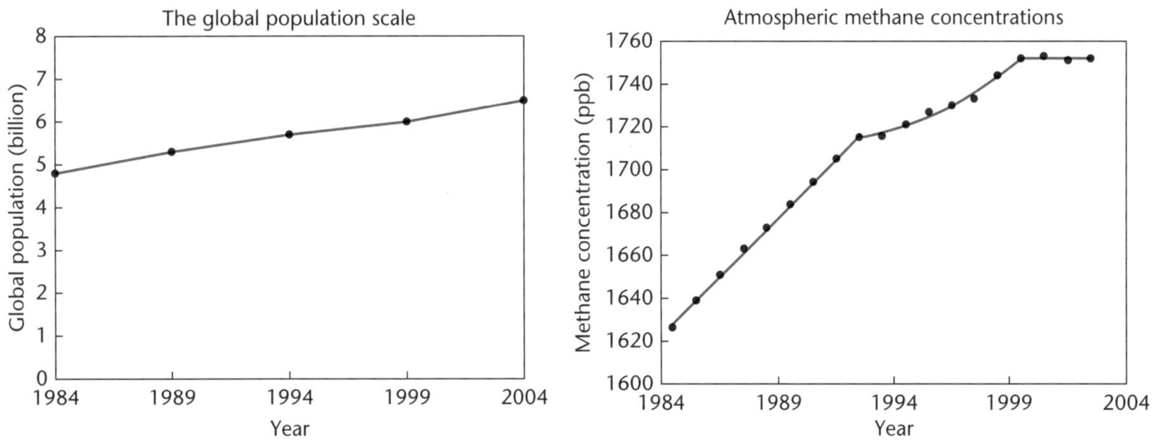

The global population scale

Atmospheric methane concentrations

Maths **a** Describe what you can conclude from the two graphs.

_____

_____

_____ [3 marks]

b Suggest why it might be difficult to reduce the methane concentration in the atmosphere.

____

______________________________ [2 marks]

Command words

To 'suggest why' means you need to apply your knowledge and understanding to a new situation.

Air pollution from burning fuels

1. Most sources of coal contain some hydrogen and sulfur, as well as carbon. Write down **three** unwanted pollutants that may be formed when coal is burned.

____

______________________________ [3 marks]

2. The pollutant nitrogen oxide is formed at high temperatures in car engines.

a Write a balanced equation, including state symbols, for its formation.

______________________________ [3 marks]

Remember

Count the atoms on each side of the equation to balance it. You can only change the number of molecules not the number of individual atoms.

b Describe how the motor industry is trying to reduce the levels of nitrogen oxides.

______________________________ [1 mark]

3. Sulfur dioxide causes acid rain.

a Write a symbol equation to show how sulfur dioxide is formed.

______________________________ [2 marks]

b Describe how acid rain damages the natural environment.

____

______________________________ [2 marks]

4. Complete the equation for the incomplete combustion of methane.

$__CH_4 + __O_2 \rightarrow __CO + __H_2O$ [2 marks]

What does the Earth provide for us?

1. The pie chart shows the sources of energy used in the UK.

Maths   **a**   Calculate the percentage of energy that comes from finite resources.

_____

_____ [2 marks]

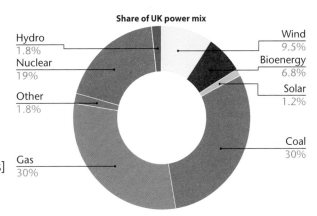

Share of UK power mix

Hydro 1.8%
Nuclear 19%
Other 1.8%
Gas 30%
Wind 9.5%
Bioenergy 6.8%
Solar 1.2%
Coal 30%

b   Describe how your answer to part (a) might change over the next 100 years. Give one **reason**.

_____

_____ [2 marks]

2. Many natural fabrics, such as cotton used to make clothes, have been replaced by synthetic fabrics such as polyester.

The table shows data for the production and weaving of 1 kg of the two fabrics.

Component	Polyester per kg	Cotton per kg
Total energy use in MJ	130	100
Oil in kg	1.5	0
Fertilisers in g	0	457
Pesticides in g	0	16
Water in litres	1308	26 100
Carbon dioxide (kg)	3.8	5.3
Approximate cost (£)	0.78	1.13

Maths   **a**   Use the data provided to write down which fabric is more sustainable to produce – polyester or cotton? Give reasons for your answer.

_____

_____ [2 marks]

b When looking at the sustainability of a product, what other factors need to be taken into account?

_____

_____ [2 marks]

3. Limestone (calcium carbonate, $CaCO_3$) is a natural material that is used to make cement and calcium oxide, CaO. During the reaction it is heated to a high temperature for long periods of time.

a Write a symbol equation for the reaction.

_____ [2 marks]

> **Remember**
> Include state symbols when you are asked to write equations.

b Evaluate the sustainability of cement.

_____

_____

_____

_____ [5 marks]

> **Command words**
> 'Evaluate' means to use the information supplied, as well as your knowledge and understanding, to consider evidence for and against.

Safe drinking water

1. Josh was provided with two samples of water and was asked to purify them. Sample **A** was seawater; sample **B** came from a lake.

Required practical

a Josh decides to purify sample **A** by distilling it. Draw a labelled diagram of the apparatus he should use.

[3 marks]

b Describe how distillation works.

_____

_____ [2 marks]

> **Practical**
> You need to be familiar with the methods used to purify water samples from different sources.

c On close observation Josh notices that sample **B** looks a bit cloudy. Describe how he could purify it.

_____

_____

_____   [4 marks]

d Once Josh had collected the purified samples, describe how he could check their purity.

_____

_____   [2 marks]

2. Describe the difference between pure water and potable water.

_____

_____   [3 marks]

3. Explain why sterilisation is the last stage of potable water production.

_____

_____   [2 marks]

Waste water treatment

1. The diagram shows how a septic tank works.

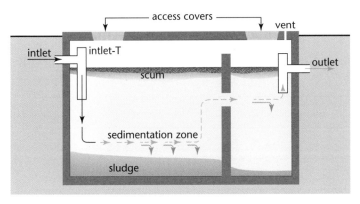

a Explain why some rural communities need septic tanks.

_____

_____ [2 marks]

b Explain why a septic tank needs an access cover.

_____ [1 mark]

c Describe what happens to the water once it has been treated.

_____

_____ [2 marks]

2. Urban lifestyles, industrial processes and rural communities produce a large amount of waste water. The diagram shows a flow diagram of the steps involved in sewage treatment.

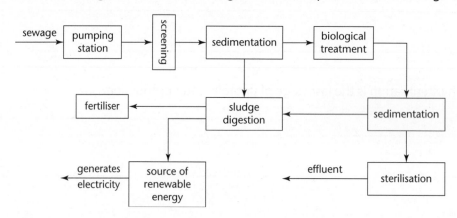

a Explain why waste water must be treated before going back into the environment.

_____

_____ [2 marks]

b There are several stages in the treatment. Explain what happens at each stage.

Screening: _____ [1 mark]

Sedimentation: _____ [1 mark]

Biological treatment: _____ [1 mark]

Sterilisation: _____ [1 mark]

c Describe how the processes involving microorganisms in the biological treatment tank differ from those in the sludge tank.

_____

_____ [2 marks]

d Give **two** uses of the sewage sludge.

_____

_____ [2 marks]

Alternative methods of extracting metals

1.

Higher Tier only

a Phytomining is an alternative way of extracting metals such as copper from low grade ores. Describe the process.

_____

_____ [2 marks]

b Explain why copper oxide is found in the ash. Write an ionic equation in your answer.

_____

_____ [2 marks]

c The copper metal is extracted from copper oxide. During the first stage of the process the copper oxide is mixed with dilute sulfuric acid. Complete the equation for this reaction.

$CuO(s) + H_2SO_4(aq) \rightarrow$ _____ + _____ [2 marks]

d Describe the final stages of this process.

_____

_____ [3 marks]

2.

Higher Tier only

Useful metals such as copper have been mined since ancient times. About 20% of the copper produced around the world now comes from bioleaching. Evaluate these alternative methods of extraction that are now being developed.

_____

_____

_____

_____

_____ [6 marks]

Life cycle assessment

1. **a** Explain why manufacturers must carry out a life cycle assessment on their products.

_____ [1 mark]

b Suggest why life cycle assessment is sometimes called 'cradle to grave analysis'.

_____

_____

_____ [3 marks]

Analysing questions

Read the question carefully. This one is worth 3 marks, so you will need to make **three** points to gain full marks.

2. Window frames can be made from wood, aluminium or plastic. An advertisement for plastic windows says *'uPVC frames have less environmental impact than other materials because they do not need maintaining or painting.'*

Evaluate whether this claim is misleading.

_____

_____

_____

_____

_____ [4 marks]

3. The table shows some life cycle assessment data for both plastic cups and paper cups.

Stage	Plastic cups	Paper cups
Raw materials	Crude oil and natural gas	Wood or wood fibre
Manufacturing	Fractional distillation: 800 °C Cracking: 200 °C and 200 atm Polymerisation: high temperature and pressure	Forestry Logs to wood pulp Beating Squeezed through rollers Drying Chemicals used for bleaching and dying
Use	Reused until they break	Usually used only once
Disposal	Landfill Non-biodegradable Recyclable	Landfill Biodegradable Recyclable

Evaluate which type of cup is more sustainable: a cup made from PET (poly(ethene) terephthalate) or a cup made from paper.

_____

_____

_____

Literacy
Use the information in the table and your knowledge and understanding of life cycle assessments to answer this question.

_____

_____

_____ [6 marks]

Ways of reducing the use of resources

1. Recycling aluminium requires less energy than is needed to extract it from its ore. Explain why.

_____

_____ [2 marks]

2. The table shows the number of plastic bags (in billions) given out at supermarkets in Britain.

Area	2010	2011	2012	2013	2014	2013/2014 change	2010/2014 change
UK	7.57	7.98	8.08	8.34	8.50	+2.3%	+12.7%
Scotland	0.75	0.75	0.76	0.80	0.65	−18.3%	−12.8%
England	6.29	6.76	7.06	7.40	7.64	+3.2%	+21.4%
Wales	0.35	0.27	0.06	0.07	0.08	+5.2%	−78.2%
Northern Ireland	0.17	0.19	0.19	0.06	0.03	−42.6%	−81.2%

a Describe the trend in plastic bag usage between 2010 and 2013.

_____ [1 mark]

b Describe how this trend changed in 2014. Suggest a reason for your answer.

_____

_____

_____ [3 marks]

Maths **c** In 2015, the 5p plastic bag charge was introduced in England. The use of plastic bags dropped by 85% compared with 2014. Calculate the number of bags used in England in 2015.

> **Maths**
> First you need to work out the percentage of bags used.

_____ [2 marks]

d Describe the environmental impact of the reduction.

_____

_____

_____ [3 marks]

Corrosion and its prevention

1. Many iron objects go rusty.

a Explain how you would prevent a bicycle chain from rusting.

_____ [2 marks]

b Magnesium blocks are often used at the seaside to prevent piers from rusting. Suggest a reason why.

_____

_____ [2 marks]

c Suggest what happens to the magnesium blocks over time.

_____ [1 mark]

d The magnesium atoms in the blocks are oxidised to Mg^{2+} ions, which causes the Fe^{3+} ions to be reduced back to Fe. Complete the equations for these reactions.

$Mg \rightarrow \_\_e^- + Mg^{2+}$ and $Fe^{3+} + \_\_\_e^- \rightarrow Fe$ [2 marks]

e Explain why magnesium is more effective at protecting iron than zinc is.

_____ [1 mark]

2. Zoe and Natasha think that both oxygen and water are needed for rusting to occur. Explain in detail how they could prove their theory.

_____

_____

_____

_____ [4 marks]

Remember
A good way to prove some scientific theories is to set up an experiment and then analyse the results.

Alloys as useful materials

1. The graph shows the effect of aluminium on the strength of copper–aluminium alloys.

Maths

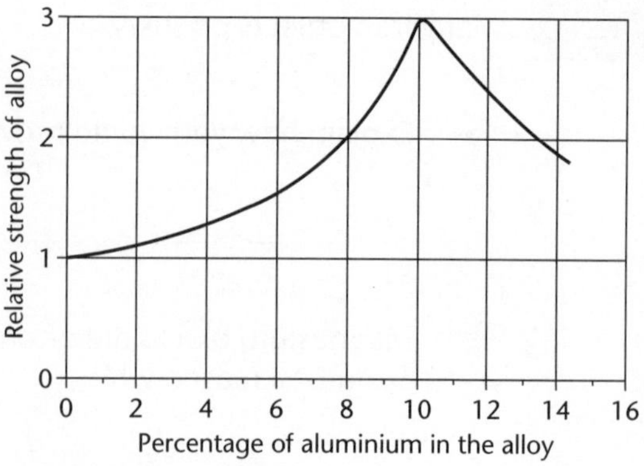

a What percentage of aluminium produces the strongest alloy? _____

[1 mark]

b What percentage of aluminium produces an alloy twice as strong as pure copper?

_____

[1 mark]

c Suggest a reason why the strength of the alloy increased at first and then decreased.

_____

_____

_____

[3 marks]

2. The image shows a spearhead made from an alloy called bronze.

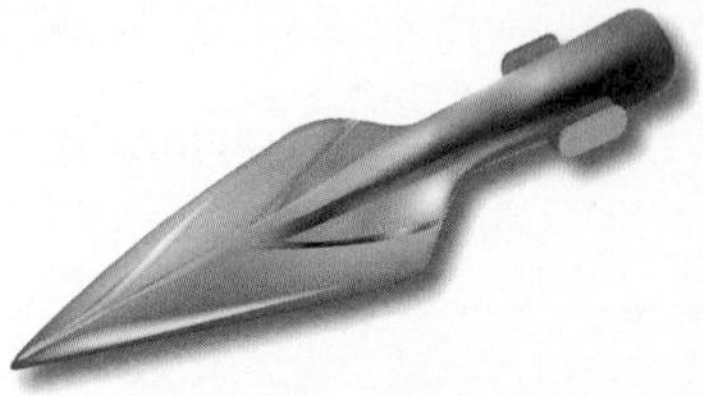

It was found that bronze made better spearheads than pure copper because it is harder than copper. Chemical analysis shows that this particular bronze alloy is 88% copper and 12% tin.

Maths **a** This spearhead weighs 1.3 kg. What mass of copper does it contain? Give your answer to 3 significant figures.

_____

_____

[2 marks]

b Explain why bronze is harder than pure copper.

_____

_____

_____

[4 marks]

Literacy

Your answer could include details of how the atoms are actually arranged in both pure copper and bronze, and why copper is relatively soft compared to bronze.

Ceramics, polymers and composites

1. **a** What type of material is concrete? Tick **one** box.

☐ Polymer   ☐ Metal   ☐ Composite   ☐ Ceramic   [1 mark]

b Tensile strength is a measure of how much force is needed to pull something apart. Suggest a reason why concrete is often reinforced using steel rods.

_____

_____

_____   [2 marks]

2. The table shows the melting points of three polymers.

Polymer	Melting point in °C	Flexibility
Low density poly(ethene)	80	Very flexible
High density poly(ethene)	120–180	Flexible
Melamine resin	331	Rigid

a Suggest which polymer you would use for a kitchen work surface. Give reasons for your answer.

_____

_____   [2 marks]

b Describe how the structure of melamine might differ from that of high-density poly(ethene).

_____

_____   [2 marks]

3. These water bottles are made from thermosoftening plastic.

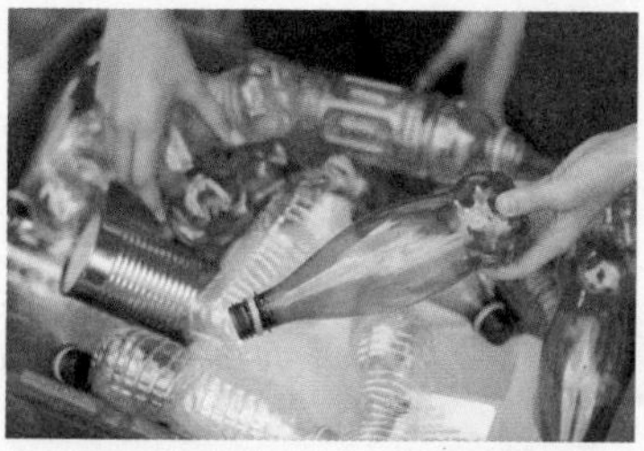

When the water has gone the bottles can be recycled. Describe and explain how these plastic bottles can be changed into new objects.

Literacy
Use ideas about polymer chains in your answer.

____

____

____

______________________________ [4 marks]

The Haber process

1. Ammonia is made in the Haber process by passing nitrogen and hydrogen over an iron catalyst. An exothermic reaction occurs.

a Write a balanced equation for the reaction. [3 marks]

________ + ________ $\rightleftharpoons$ ________

b Give the meaning of the symbol $\rightleftharpoons$

______________________________ [1 mark]

Maths

Higher Tier only

The graph shows how the percentage of ammonia produced changes with temperature and pressure.

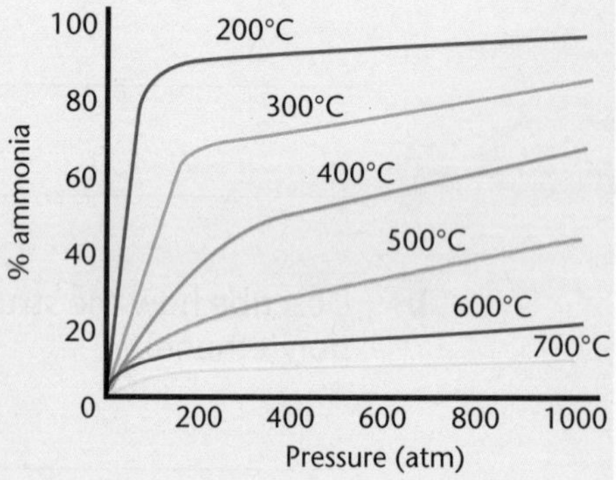

c Describe how the percentage of ammonia produced changes with pressure.

______________________________ [1 mark]

d Explain your observation.

____

______________________________ [2 marks]

e Describe how the percentage of ammonia produced changes with temperature.

______________________________ [1 mark]

f Explain your observation.

_____

_____ [2 marks]

Synoptic **g** Industrial ammonia is produced in large quantities using these conditions:

- Temperature 400–450 °C.

- Pressure 200 atm.

- Iron catalyst.

Explain why these conditions are chosen.

> **Synoptic**
>
> To answer this question use the information you have been given, along with your knowledge of equilibrium from Section 6.

_____

_____

_____

_____ [4 marks]

Production and use of NPK fertilisers

....

1. Potassium nitrate is an important fertiliser with an N–P–K ratio of 13.7 : 0 : 46.2

a Explain what is meant by the term 'NPK ratio'. [2 marks]

It is the ratio/percentage of nitrogen : phosphorus : potassium compounds in the fertiliser. [2 marks]

Comment: This answer is correct.

b Explain why farmers use a range of different fertilisers. [3 marks]

Different crops take different elements out of the soil, which need replacing by fertilisers. [1 mark]

Comment: This answer does not explain why different fertilisers are needed. It's a good start but it needs to go on to say that different fertilisers have different NPK ratios. So, for example, potassium nitrate is good for a soil low in potassium but will not help a soil that needs phosphorus because it doesn't contain any.

c Describe how potassium nitrate is made in the laboratory. Write an equation for the reaction. [4 marks]

The equation is for the reaction is:

nitric acid + potassium hydroxide → potassium nitrate + water

Method

- Measure the potassium hydroxide solution using a measuring cylinder and pour it into a conical flask.

- Use a burette to add nitric acid to the flask slowly until the mixture is neutral. You will need to use a pH probe to show this.

- Pour the solution into an evaporating basin and heat gently until crystals form.

- Filter off the crystals using a filter funnel and paper. Leave the crystals to dry.

Comment: This is a good answer. It starts with the word equation. Only write a symbol equation if you are confident that you will get it right. Each step of the method is given using bullet points, which makes it easy to follow.

Marks gained: [4 marks]

2. Compare the sustainability and the impact on the environment of the processes used to manufacture ammonia and fertilisers. Use these words in your answer.

Synoptic **non-renewable** **resources** **transport** **fossil fuels** **Haber process** **mining**

_____

_____

_____

_____

> **Synoptic**
> To answer this question, you will need to use your knowledge and understanding of industrial processes as well as ideas about the environmental impact of the processes involved.

_____

_____

_____ [6 marks]

Answers

Section 1: Atomic structure and the periodic table

Atoms, elements and compounds

1. Elements are made up of only one type of atom. [1 mark] Compounds contain two or more elements [1 mark] that are combined chemically. [1 mark]

2. **a** O_2 [1 mark]; Hg [1 mark]

   b It separates into elements [1 mark]; when it is heated. [1 mark]

   c Mercury oxide [1 mark]

Mixtures

1. **a** Distillation/simple distillation [1 mark]

   b Pour the mixture into the flask and heat it [1 mark]. When the mixture reaches 78 °C the ethanol will start to boil and turn into a gas [1 mark]. The ethanol gas will condense in the tube and turn into a liquid (which is collected in the beaker) [1 mark]. Stop heating before the mixture reaches a temperature of 100 °C when the water boils. [1 mark]

Compounds, formulae and equations

1. $FePO_4$ [1 mark]

2. **a** Sodium nitrate; [1 mark] **b** Potassium hydroxide; [1 mark]; **c** Calcium carbonate; [1 mark] **d** Magnesium sulfate. [1 mark]

3. zinc + hydrochloric acid → zinc chloride + hydrogen [1 mark]
   Zn [1 mark] (s) [1 mark] + 2HCl (aq) → $ZnCl_2$(aq) + H_2 [1 mark] (g) [1 mark]

Scientific models of the atom

1. Positively charged ball; [1 mark] with negatively charged particles/electrons embedded in it. [1 mark]

2.
Level 3: An explanation of what the results show using the correct terminology. Shows a clear understanding of how new evidence led to changes in models and theories. [5–6 marks]
Level 2: Uses the results to explain why the model had to change. [3–4 marks]
Level 1: Gives one or more key points [1–2 marks]
Indicative content

   - The prediction was based on the original (plum pudding) model.
   - New evidence was collected.
   - The results showed that the (plum pudding) model was not correct.
   - The model had to change to explain the results.
   - The model was changed to include a central, positively charged nucleus.

Sizes of atoms and molecules

2. **a** 2×10^{-10} [1 mark]

   b 0.2/0.00002 = 10 000 times [1 mark]

   c 150 ÷ 10 000 [1 mark] = 0.015 m [1 mark]; 0.015 × 1000 = 15 (mm) [1 mark] (allow mistake from **b** to be carried over)

Relative masses and charges of subatomic particles

1. They contain the same number of protons and electrons: [1 mark]; so the negative and positive charges cancel each other out. [1 mark]

2. **a** 4 [1mark]; **b** 9 [1 mark]; **c** Be [1 mark]

3. Both contain 17 protons. [1 mark] Both contain 17 electrons. [1 mark]
   Chlorine-35 contains 18 neutrons; [1 mark] chlorine-37 contains 20 neutrons. [1 mark]

Relative atomic mass

1. **a** There should be 7 particles altogether [1 mark]; 3 with plus symbols. [1 mark]

   b Lithium atoms are mainly lithium-7; [1 mark] because the relative atomic mass is nearly 7. [1 mark]

2. Mg_{24} Mg_{25} Mg_{26}
   79 × 24 = 1896   10 × 25 = 250   11 x 26 = 286 [1 mark]
   (1896 + 250 + 286) ÷ 100 [1 mark]
   = 24.32 [1 mark]

Electronic structure

1. 2,7 [1 mark]

2. 2 electrons in innermost energy level; [1 mark] 8 electrons in second energy level; [1 mark]; 8 electrons in third energy level. [1 mark] Charge −1. [1 mark] It has one more electron than protons/17 protons and 18 electrons [1 mark] because protons have a positive charge and electrons have a negative charge. [1 mark]

Electronic structure and the periodic table

2. **a** $2Li + Cl_2$ [1 mark] → 2LiCl [1 mark]

   b The glow will be more intense/a flame will be produced. [1 mark] The reaction will be over more quickly. [1 mark] Sodium is more reactive than lithium. [1 mark] A white powder/sodium chloride is produced. [1 mark]

Development of the periodic table

1. **a** A_r for lithium = 7; A_r for potassium = 39. 7 + 39 = 46 [1 mark]; 46 ÷ 2 = 23 [1 mark] and this is the relative atomic mass of sodium. [1 mark]

   b Any one from: new elements were discovered that didn't fit (into triads); Döbereiner's arrangement didn't apply to all elements. [1 mark]

2. It is a metal. [1 mark] It has a melting point lower than 660 °C (any reasonable estimation). [1 mark] It has a density higher than 2.70 g/cm³ (any reasonable estimation). [1 mark] It reacts with chlorine to form a compound with the formula XCl_3. [1 mark]

Comparing metals and non-metals

1. **a** Oxygen [1 mark]

    b It will gain 2 electrons when it reacts [1 mark] to form a negative (−2) ion. [1 mark]

2. **A** has a lower density than most metals. [1 mark]
    B is a non-metal that conducts electricity. [1 mark]
    C is a metal with a low melting point/liquid at room temperature. [1 mark]

Elements in Group 0

1. **a** All points plotted correctly. [2 marks] Just 1 mark if one point is plotted incorrectly. Line of best fit drawn. [1 mark]

    b The answer given should correspond to the drawn line of best fit. [1 mark]

2. They have a full outer shell of electrons. [1 mark] So do not have to lose or gain any electrons (to gain a full outer shell). [1 mark] So do not react/form compounds with other elements. [1 mark]

Elements in Group 1

1. $4Li + O_2$ [1 mark] $\rightarrow 2 Li_2O$ [1 mark] Correct balancing [1 mark]

2. **a** Volume = $2.5 \times 5.0 \times 1.8 = 22.5$ (cm³) [1 mark]
    Mass = $22.5 \times 0.53 = 11.93$ (g) [1 mark]

    b The melting point decreases. [1 mark] The density increases; [1 mark] But potassium has a lower density than sodium/potassium is an anomaly. [1 mark]

Elements in Group 7

2. **a** The more reactive chlorine [1 mark] has displaced the less reactive iodine from the compound. [1 mark]

    b $2I^-(aq) + Cl_2(g)$ [1 mark] $\rightarrow 2Cl^-(aq) + I_2(aq)$ [1 mark]

3. When a Group 7 element reacts with a metal it gains an electron. [1 mark] To form a stable arrangement/to fill its outer shell of electrons/to get 8 electrons in its outer shell. [1 mark] The electron is attracted to the positive nucleus. [1 mark] As you go down the group the atom get larger. [1 mark] The outer shell is further from the nucleus/nucleus is shielded by more electron shells. [1 mark] The larger the atom, the weaker the force of attraction from the nucleus on the electron. [1 mark]

Properties of the transition metals

1. **a** Any two from: high melting point; malleable; strong; hard for [1 mark] each.

    b Any one of: does not rust; corrode; react with water for [1 mark]

2. **a** Cobalt [1 mark]; **b** +3 [1 mark]

3.
Level 3: Compares at least two physical properties and one chemical property of Group 1 metals and the transition metals. The answer is detailed and describes how the properties are different. The answer may contain specific examples or reactions. [5–6 marks]

Level 2: Compares at least two physical properties and one chemical property of Group 1 metals and the transition metals. [3–4 marks]
Level 1: One or more key points are given [1–2 marks]

Indicative content
Physical properties:
*   Transition metals are harder/more dense/stronger than Group 1 metals.
*   Transition metals have higher melting/boiling points than Group 1 metals.
*   Physical properties:

Chemical properties:
*   Group 1 metals are more reactive than transition metals.
*   All Group 1 metals react quickly with water/oxygen/halogens.
*   Transition metals react slowly (e.g. zinc/iron) or not at all (e.g. gold) with oxygen/water.

Section 2: Bonding, structure and the properties of matter

The three states of matter

1. **a** Wait until the salol has turned into a liquid. [1 mark] Read the temperature using the thermometer. [1 mark]

    b The forces of attraction between the particles in magnesium oxide are stronger than those in salol. [1 mark] So more heat energy is needed to overcome them. [1 mark]

2.
Level 3: Mentions more than one argument for and against the model. [5–6 marks]
Level 2: Mentions at least one argument for and against. [3–4 marks]
Level 1: Mentions one argument for or against. [1–2 marks]

Indicative content
Arguments for:
*   It shows how the particles are arranged in each state.
*   It shows that the particles are identical in each state.

Arguments against:
*   There are no forces shown between the particles.
*   It does not show the movement of the particles.
*   All particles are represented as spheres.
*   The spheres are solid.

Ionic bonding and ionic compounds

1. **a** Cu^{2+} and SO_4^{2-} [1 mark]

    b It is made up of a metal and a non-metal. [1 mark]

2. **a** A [1 mark]

Dot and cross diagrams for ionic compounds

2. Outer shell of electrons on atoms drawn correctly (2 for magnesium, 7 for chlorine). [1 mark] Two electrons transferred from the magnesium atom [1 mark] forming Mg^{2+}/2+ ions. [1 mark] Two chlorine atoms gain 1 electron each [1 mark] forming Cl^-/ 1− ions. [1 mark] The compound formed is magnesium chloride. [1 mark] and has the formula $MgCl_2$ [1 mark]

Properties of ionic compounds

1. **a** When solid the ions cannot move. [1 mark] When melted the ions are free to move. [1 mark]

   b There are strong forces of attraction between the ions. [1 mark] This comes about because of the attraction between positive and negative ions. [1 mark] Large amounts of energy are needed to overcome the forces. [1 mark] There are stronger forces between calcium ions and oxygen ions (than between the sodium and chloride ions). [1 mark] In calcium oxide two electrons are transferred to each oxygen atom; in sodium chloride only 1 electron is transferred to each chloride atom; [1 mark] (accept an answer mentioning the different charges on the ions). The more electrons transferred (in each ionic bond), the stronger the force of attraction. [1 mark]

Covalent bonding in small molecules

1. CO_2 [1 mark]; Cl_2 [1 mark]

2. **a** Water [1 mark]

   b By sharing electrons. [1 mark]

3. **a** Intermolecular (force)/covalent bond [1 mark]

   b Same molecules representing the gaseous phase. [1 mark] Separated with space between. [1 mark]

   c It is lower; [1 mark] because the forces of attraction between the molecules of the substance are lower than those between the (charged) ions in sodium chloride. [1 mark] So less energy is needed to overcome them. [1 mark]

Dot and cross diagrams for covalent compounds

1. They do not show the 3D shape of molecules. [1 mark]

2.

   Five electrons on the outer shell of the nitrogen atom. [1 mark] One electron in each outer shell of the hydrogen atoms. [1 mark] One shared pair of electrons between the nitrogen atom and each hydrogen atom. [1 mark] One unshared pair of electrons remaining on the nitrogen atom. [1 mark]

Properties of small molecule compounds

1. **a** Covalent (bond). [1 mark] **b** Intermolecular. [1 mark]

2. They have weak (intermolecular) forces between the molecules, [1 mark] which do not require much (heat) energy to overcome. [1 mark]

3. As the molecules get bigger, the boiling points increase. [1 mark] The larger the molecule, the stronger the forces of attraction between the molecules; [1 mark] so more heat energy is needed to overcome the forces; [1 mark] so the particles can move apart and form a gas. [1 mark]

Polymers

1. (Square) brackets drawn around repeating unit; [1 mark] 'n' added bottom of right bracket. [1 mark]

2. **a** Ethene is the repeating unit of poly(ethene)/ many ethene molecules joined together form poly(ethene). [1 mark]

   b The intermolecular forces are stronger in poly(ethene) (than in ethene). So more heat energy is needed to overcome the forces and turn poly(ethene) into a gas.

Giant covalent structures

1. **a** Simple molecules: water, ammonia [2 marks]; Giant covalent: diamond, silicon dioxide. [2 marks]

   b Diamond; [1 mark] ammonia; [1 mark] diamond [1 mark]

2. Any four from: Silicon dioxide has a giant covalent structure. [1 mark] Carbon dioxide is made up of simple molecules. [1 mark] There are strong covalent bonds between the silicon atoms and the oxygen atoms in silicon dioxide. [1 mark] All the bonds in silicon dioxide are strong covalent bonds. [1 mark] There are strong covalent bonds between the carbon atoms and the oxygen atoms in carbon dioxide. [1 mark] There are weak intermolecular forces between carbon dioxide molecules. [1 mark]

Properties of giant covalent structures

1. **a** Contains only one substance/is not a mixture. [1 mark]

   b For [1 mark each] any two from: High melting/ boiling point; Insoluble in water; Hard; Strong.

   c Both have strong covalent bonds between their atoms [1 mark] which require a lot of energy to break. [1 mark]

2.

Level 3: Gives a number of similarities and differences between the physical properties of diamond and graphite. Each is explained using correct scientific knowledge. [5–6 marks]
Level 2: Gives one or two similarities and differences. They are mostly explained correctly. [3–4 marks]

Level 1: Gives one or two similarities and differences. The reasons are incorrect or incomplete. [1–2 marks]

Indicative content

Similarities:

- The individual layers in graphite and diamond are both hard and strong.
- Both have a high melting point and a high boiling point.
- Both are insoluble in water.
- The covalent bonds between the carbon atoms are strong and difficult to break.

Differences:

- Graphite can conduct electricity; diamond can't.
- In graphite there are (delocalised) electrons between the layers which can move/carry charge.
- Diamond has no free electrons.
- The overall structure of graphite is weak; the overall structure of diamond is strong.
- The intermolecular forces between the layers in graphite are weak so can break with little force.

Graphene and fullerenes

1. **a** The carbon nanotube. [1 mark] It has a 3D shape/it is hollow. [1 mark]

   b They speed up/increase the rate of a chemical reaction. [1 mark] Atoms of the catalyst can be attached to a nanotube, [1 mark] increasing the surface area of the catalyst [1 mark] so there is more chance of the reactants colliding with the catalyst. [1 mark]

   c It can conduct electricity. [1 mark] This is useful because touchscreens need to be able to conduct electricity. [1 mark] It has high strength. [1 mark] This is useful so the screen does not crack/shatter if dropped. [1 mark] It is transparent. [1 mark] This is useful so you can see the light coming through from the display underneath. [2 marks]

Nanoparticles

1. **a** 2 orders. [1 mark]

   b Titanium dioxide. [1 mark] It is smaller than 100 nm. [1 mark]

2. Volume = $2 \times 2 \times 2 = 8$ cm^3; [1 mark] Surface area = $(2 \times 2) \times 6 = 24$ cm^2. [1 mark] The ratio is 3 : 1 (accept 24 : 8). [1 mark]

Uses of nanoparticles

1. **a** It increases the validity [1 mark] because other scientists have checked the research. [1 mark]

   b
Level 3: Several reasons for and against a ban are discussed with valid reasoning. [5–6 marks]
Level 2: Several reasons for and against are listed. [3–4 marks]
Level 1: One or two reasons for and against a ban are stated. [1–2 marks]

Indicative content

- Nanoparticles have many uses – e.g. applications in medicine, in electronics, in cosmetics and sun creams, as deodorants and as catalysts.
- There are many potential uses of nanoparticles that will not be discovered if they are banned.
- There might not be any risk to human health – more research is needed.
- Nanoparticles are only a risk to human health if they are inhaled, like the PM$_{2.5}$ particulates. In the uses stated they cannot be inhaled.
- PM$_{2.5}$ particles are larger than the nanoparticles used in these applications, so nanoparticles may be even more dangerous.
- Nanoparticles might pose a risk to human health, so a ban would prevent people from getting ill.
- Because the technology is so new, health problems may become apparent in the future.

Metallic bonding

1. They are not attached to any particular atom. [1 mark] They are free to move. [1 mark]

2. **a** The melting points of Group 2 metals are higher [1 mark] than the melting points of Group 1 metals.

   b Group 2 metals have 2 electrons in their outer shell/the delocalised electrons in Group 2 metals are made up of 2 electrons from each atom. [1 mark]

   Group 1 metals have 1 electron in their outer shell/the delocalised electrons in Group 1 metals are made up of 1 electron from each atom. [1 mark]

   There are more delocalised electrons in Group 2 metals. [1 mark] So the metallic bonds are stronger in Group 2 metals than in Group 1 metals, so more heat energy is needed to overcome them. [1 mark]

Properties of metals and alloys

1. Graphite has a low density because its atoms are arranged in layers with space in between. [1 mark] Iron has a high density because its atoms are packed close together. [1 mark] Iron is malleable because the layers of atoms can slide over each other. [1 mark] Graphite is brittle because there are weak forces between the layers of atoms. [1 mark] They can both conduct electricity because they have delocalised electrons. [1 mark] The delocalised electrons are free to travel through the material. [1 mark]

Section 3: Quantitative chemistry

Writing formulae

1. **a** CaCO$_3$ [1 mark]; HCl [1 mark]

   b [1 mark] for any of: Use a gas syringe; Use a delivery tube into an upturned measuring cylinder/burette full of water. Accept a labelled diagram.

 c Calcium chloride. [1 mark]

2. a Combustion (accept 'oxidation'). [1 mark]

 b 3 CO_2 molecules (1 mark); 10 oxygen atoms [1 mark]

Conservation of mass and balanced chemical equations

1. a $Pb^{2+}(aq)$ [1 mark] + $2I^-(aq)$ [1 mark] $\rightarrow$ $PbI_2(s)$ [1 mark]

 b Lead iodide. [1 mark] It is more dense than the surrounding liquid. [1 mark]

 c 3.9 − 2.4 [1 mark] = 1.5 g [1 mark]

2. $Ca(OH)_2$ + 2HCl [1 mark] $\rightarrow$ $CaCl_2$ + $2H_2O$ [1 mark]

Mass changes when a reactant or product is a gas

2. a All points plotted correctly. [2 marks] (8, 9 or 10 points scores 1 mark) Curved line of best fit [1 mark]

 b Carbon dioxide/a gas is produced [1 mark] which leaves the container. [1 mark]

Relative formula mass

2. a $C_8H_9NO_2$ [1 mark]

 b $M_r = (12 \times 8) + (1 \times 9) + (1 \times 14) + (2 \times 16)$ [1 mark] = 151 [1 mark]

3. M_r of 2KI = 2 × (39 + 127) = 332 [1 mark]
M_r of $Pb(NO_3)_2$ = 207 + (14 × 2) + (16 × 6) = 331 [1 mark]
Total M_r of reactants = 332 + 331 = 663 [1 mark]
M_r of PbI_2 = 207 + (2 × 127) = 461 [1 mark]
M_r of $2KNO_3$ = 2 × (39 + 14 + (16 × 3)) = 202 [1 mark]
Total M_r of products = 461 + 202 = 663 [1 mark]

Moles

1. a M_r of CO_2 = 12 + (16 × 2) = 44; [1 mark] Number of moles = 6.60 ÷ 44 = 0.15 [1 mark]

 b M_r = (1 × 2) + 16 = 18; [1 mark]
Mass = 18 × 0.20 = 3.6 g [1 mark]

 c $6.02 \times 10^{23} \times 0.2 = 1.204 \times 10^{23}$; [1 mark] = 1.20×10^{23} to 3 s.f. [1 mark]

Amounts of substances in equations

1. a Thermal decomposition. [1 mark]

 b M_r magnesium carbonate = 24 + 12 + (16 × 3) = 84 [1 mark]
Moles of magnesium carbonate = 4.2 ÷ 84 = 0.05 [1 mark]
M_r magnesium oxide = 24 + 16 = 40 [1 mark]
Mass of magnesium oxide = 40 × 0.05 = 2.0 (g) [1 mark]

 c For [1 mark] any from: They did not all heat the magnesium carbonate for long enough/in a hot enough flame. Some did not measure the mass of magnesium carbonate/magnesium oxide correctly.

 d Range: 3.7 − 2.1 = 1.6 g [1 mark]
Mean: (2.4 + 2.2 + 3.7 + 2.9 + 2.1 + 2.7 + 3.1 + 3.2) ÷ 8 = 2.7875 g [1 mark]
Percentage uncertainty = (1.6 ÷ 2.7875) × 100 [1 mark] = 57.4% [1 mark]

Using moles to balance equations

1. M_r calcium carbonate = 40 + 12 + (16 × 3) = 100; [1 mark] Number of moles produced = 200 ÷ 100 = 2(moles) [1 mark]

2. a Iron + chlorine $\rightarrow$ iron chloride [1 mark]

 b Number of moles of iron (Fe) = 1.12 ÷ 56 [1 mark] = 0.02; [1 mark] Number of moles of chlorine (Cl_2) = 2.13 ÷ (35.5 × 2) [1 mark] = 0.03 [1 mark]

 c 2Fe + $3Cl_2$ [1 mark] $\rightarrow$ $2FeCl_3$ [1 mark]

Limiting reactants

1 a 2K + $2H_2O$ [1 mark] $\rightarrow$ 2KOH + H_2 [1 mark]

 b Potassium. [1 mark] There is none left at the end of the reaction. [1 mark]

2. a There would be some copper oxide (in the beaker) after stirring. [1 mark]

 b To make sure all the acid is reacted. [1 mark] To increase the yield of copper sulfate. [1 mark]

 c M_r CuO = 63.5 + 16 = 79.5 [1 mark]; Mass of CuO = 79.5 × 0.05 = 3.975 [1 mark] = 4 g (to 1 s.f.) [1 mark]

Concentration of solutions

1. 1 g of salt in 2 cm^3 of water [1 mark]

2. a Concentration = mass ÷ volume = 22 ÷ 100 = 0.22 (g/cm^3) [1 mark]

 b Volume = 1.5 × 1000 = 1500 cm^3; [1 mark] Concentration = 450 ÷ 1500 = 0.3 (g/cm^3) [1 mark]

 c Volume = 75 cm^3 ÷ 1000 = 0.075 dm^3 [1 mark]; mass = volume × concentration = 0.075 × 20 = 1.5 g [1 mark]

3. The concentration increases. [1 mark] The water/solvent will evaporate [1 mark] and the volume of the solvent/water will decrease. [1 mark] The mass of the salt will stay the same/the salt will not evaporate. [1 mark]

Using concentrations of solutions in mol/dm^3

1. a M_r $Cu(NO_3)_2$ = 63.5 + (14 × 2) + (16 × 6) = 187.5 [1 mark]
Mass of 0.5 moles = 0.5 × 187.5 = 93.5; [1 mark] Add 140.625 g (93.75 x 1.5) of copper(II) nitrate [1 mark] to 1.5 dm^3 of water. [1 mark]

 b Moles of $Cu(NO_3)_2$ = 0.5 mol/dm^3 × (20 ÷ 1000) [1 mark] = 0.01 [1 mark]
1 mole of $Cu(NO_3)_2$ reacts with 2 moles of NaOH. [1 mark]
So the amount of NaOH that reacts is 0.01 × 2 = 0.02 moles. [1 mark]
Its concentration is 0.02 ÷ (45.3 ÷ 1000) = 0.4 mol/dm^3. [1 mark]

Amounts of substances in volumes of gases

1. Number of moles of CO_2 = 220 ÷ (12 + (16 × 2)) = 5. [1 mark] So the gas volume = 5 × 24 dm^3 = 120 dm^3. [1 mark]

2. a The mass will decrease [1 mark] because hydrogen is being produced and then escapes into the air. [1 mark]

 b Number of moles of zinc = 2 ÷ 65 = 0.03 = number of moles of hydrogen. [1 mark] Volume = 0.03 × 24 = 0.7 dm³ [1 mark]

3. Methane [1 mark]

Number of moles of the hydrocarbon = 4800 ÷ 24000 = 0.2 [1 mark]

M_r of the hydrocarbon: 3.2 ÷ 0.2 = 16 [1 mark]

M_r of methane = 12 + (1 × 4) = 16 [1 mark]

Percentage yield

2. a Either: the fizzing/bubbling will stop; or the pH will be 7/neutral. [1 mark]

 b Moles of ethanoic acid = 0.5 × (25 ÷ 1000) = 0.0125 [1 mark]

2 moles of ethanoic acid = 1 mole of magnesium ethanoate. [1 mark]

M_r magnesium ethanoate = ((12 + 3 + 12 + 16 + 16) × 2) + 24 = 142 [1 mark]

Mass of magnesium ethanoate = 142 × (0.0125/2) = 0.8875g [1 mark]

Atom economy

1. It is important for sustainable development. [1 mark] To increase the amount of money they make. [1 mark]

2. a M_r of aspirin: (12 × 9) + 8 + (16 × 4) = 180 [1 mark]
M_r of all reactants = (12 × 7) + 6 + (16 × 3) + (12 × 4) + 6 + (16 × 3) = 240 [1 mark]
(180 ÷ 240) × 100 = 75% [1 mark]

 b. For [1 mark] any one from: It gives a lower yield of aspirin; the reactant ethanoic acid is more expensive than ethanoic anhydride; the first reaction makes ethanoic acid which is a useful product; this increases the profits made by the drug company.

Section 4: Chemical changes

Metal oxides

1. a Oxidation; [1 mark] **b** Reduction; [1 mark]
 c Oxidation [1 mark]

2. a Heat the zinc [1 mark] in air/oxygen. [1 mark]
$2Zn + O_2$ [1 mark] $\rightarrow 2ZnO$ [1 mark]

 b Oxygen has joined/combined with the zinc. [1 mark]

 c (350 ÷ 373.8) × 100 = 93.6% [1 mark]

Reactivity series

1. Sodium loses electrons more easily. [1 mark]

2. a Add each metal to some acid and also to some water. [1 mark] Hydrogen gas/bubbles will be seen during the reactions. [1 mark] The calcium will react quickly in both. [1 mark] Magnesium will react in

both, but less quickly than calcium. [1 mark] Zinc will react slowly in acid, but very slowly in water. [1 mark] Silver will not react in either acid or water. [1 mark]

 b It is very reactive [1 mark], so it would not be safe [1 mark].

Reactivity series – displacement

1. Iron [1 mark]

2. Add the metals to the copper sulfate solution. [1 mark]. Only zinc will react [1 mark] because zinc is more reactive than copper. [1 mark]

Extraction of metals

1. a For 1 mark each: Iron and lead to 'reduction with carbon'. [2 marks] Magnesium and sodium to 'electrolysis'.

 b It is found pure; [1 mark] because it is very unreactive; [1 mark] and does not react with other elements/does not form compounds. [1 mark]

2. a Zinc oxide [1 mark]

 b Carbon [1 mark]

 c Quarrying the oxide/ore [1 mark] can destroy habitats/reduce biodiversity [1 mark]; or Carbon dioxide is produced, which is a greenhouse gas [1 mark] and contributes to climate change [1 mark]; or Carbon/zinc oxide/ores are limited resources [1 mark] so supplies will run out sometime. [1 mark]

Oxidation and reduction in terms of electrons

1. A non-metal gains electrons to form ions. [1 mark]

2. a $Fe(s) + Cu^{2+}(aq)$ [1 mark] $\rightarrow Fe^{2+}(aq) + Cu(s)$ [1 mark]

 b Copper [1 mark]; Iron [1 mark]

3. Reduction; [1 mark] because hydrogen ions are gaining electrons. [1 mark]

Reactions of acids with metals

1. Hydrogen; [1 mark] Zinc sulfate. [1 mark]

2. 2 [1 mark] and $MgCl_2$ [1 mark]

3. a Iron sulfate [1 mark]

 b Iron; [1 mark] Hydrogen [1 mark]

Neutralisation of acids and making salts

1. It is insoluble in water. [1 mark] It neutralises acids. [1 mark]

2. a Acid: hydrochloric (acid); [1 mark] Alkali: potassium hydroxide

 b KCl [1 mark]

3. a For [1 mark] each: $CuSO_4$, H_2O and CO_2.

 b It will go cloudy [1 mark] because carbon dioxide is produced. [1 mark]

Making soluble salts

1. **a** Zinc; [1 mark] Hydrochloric (acid) [1 mark]

   b Put a lighted splint into the gas. [1 mark] There will be a squeaky pop. [1 mark]

2.

Level 3: A clear, detailed description of the method, which names the equipment needed correctly and explains the purpose of each step. There is a comprehensive list of appropriate safety precautions. [5–6 marks]
Level 2: There is a clear description of the method, which includes most of the equipment needed, and an explanation of the steps in the procedure. Some of the safety precautions are included. [3–4 marks]
Level 1: A basic method that includes some of the equipment, and there is some attempt to explain some of the steps. [1–2 marks]

   Indicative content

   - Heat the sulfuric acid in a beaker using a Bunsen burner.
   - Add the copper oxide while stirring.
   - Heating and stirring speeds up the reaction.
   - Add copper oxide until it is in excess, which means that the acid is fully neutralised.
   - Filter the mixture by pouring it through a funnel and filter paper (accept a labelled diagram).
   - Filtering removes excess/unreacted copper oxide.
   - Pour the solution into an evaporating basin.
   - Heat it gently using a water bath or electric heater to evaporate the water.
   - Leave the remaining solution to crystallise the salt.

pH and neutralisation

1. 7 with neutral; [1 mark] 8–14 with alkaline; [1 mark] 1–6 with acidic [1 mark]

2. OH^-; [1 mark] H_2O [1 mark]

3. **a** Add some universal indicator to the sodium hydroxide solution. Add small amounts of acid, swirling after each addition, until the indicator just turns green and stays green.

   b It is difficult to tell when the solution is just neutral because there is no precise colour change.

   c You can use a pH probe instead.

Titrations

1. **a** Add acid from the burette a drop at a time. [1 mark] Swirl after each addition. [1 mark] Until the colour of the indicator just changes. [1 mark] Repeat [1 mark]

   b To see colour change of indicator clearly. [1 mark] To get an accurate volume of acid (needed to neutralise the alkali). [1 mark]

2. Moles of $H_2SO_4 = (22.30 \div 1000) \times 0.100$ [1 mark] $= 0.00223$ [1 mark]
   Moles of NaOH $= 0.00223 \times 2 = 0.00446$ [1 mark]
   Concentration of NaOH $= 0.00446 \div (25 \div 1000)$ [1 mark] $= 0.1784$ [1 mark]
   $= 0.178$ to 3 d.p. [1 mark]

Strong and weak acids

1. **a** Acid B: Citric acid [1 mark]

   b H^+; [1 mark] Cl^- [1 mark]

2. **a** $0.000\ 001 = 10^{-6}$; [1 mark] pH = 6 [1 mark]

   b The pH will increase; [1 mark] because the concentration of hydrogen ions decreases. [1 mark]

3. It ionises fully when it forms a solution. [1 mark] So it has a high concentration of hydrogen (H^+) ions. [1 mark]

The process of electrolysis

1. So the ions are free to move. [1 mark]

2. **a** Power supply added. [1 mark] Negative electrode labelled 'cathode'. [1 mark] Positive electrode labelled 'anode'. [1 mark] Liquid labelled 'electrolyte' or 'copper chloride solution'. [1 mark]

   b Negative/cathode [1 mark] because they are positive ions. [1 mark]

3. **a** e^-; [1 mark] X [1 mark]

   b Reduction [1 mark]

Electrolysis of molten ionic compounds

1. Sodium chloride (NaCl); [1 mark] Calcium oxide (CaO) [1 mark]

2. **a** The ions (in the lead(II) bromide) are free to move; [1 mark] so it can complete the circuit; [1 mark] and conduct electricity. [1 mark]

   b Bubbles/bromine gas forms at the positive electrode/anode; [1 mark] because negative bromide ions are attracted. [1 mark] A metal forms at the negative electrode/cathode; [1 mark] because positive lead ions are attracted. [1 mark]

Using electrolysis to extract metals

1. **a** Aluminium is more reactive than carbon. [1 mark]

   b Aluminium ions have a positive charge. [1 mark] They are attracted to the (negative) cathode. [1 mark]

   c 3 [1 mark]

   d Negative oxygen ions/O^{2-} ions [1 mark] are attracted to the anodes. [1 mark] They react to form carbon dioxide. [1 mark] $C + O_{2-} \rightarrow CO_2$ [1 mark]

Electrolysis of aqueous solutions

1. **a** Sulfate ions and hydroxide ions. [1 mark]

   b Put a glowing splint into the gas. [1 mark] If it relights, oxygen is present. [1 mark]

   c Copper has been discharged. [1 mark] Copper is less reactive than hydrogen. [1 mark] $Cu^{2+} + 2e^-$ [1 mark] $\rightarrow Cu$ [1 mark]

2. Anode: Bromine; [1 mark] Cathode: Hydrogen [1 mark]

Half equations at electrodes

1. **a** **A**; [1 mark] **b** **A**; [1 mark] **c** **B** [1 mark]

2. **a** $2e^-$ [1 mark]

b $2H^+$ [1 mark] + $2e^-$ [1 mark] $\rightarrow H_2$

c Sodium hydroxide. [1 mark] Sodium (Na^+) ions and hydroxide (OH^-) ions are left over in solution. [1 mark]

3. a $2e^-$; [1 mark] **b** $4OH^-$ [1 mark]

Section 5: Energy changes in reactions

Exothermic and endothermic reactions

1. Sports injury packs with endothermic. [1 mark] Hand warmer and self-heating can with exothermic. [2 marks] Photosynthesis is not an application.

2. a Carbon dioxide [1 mark]; **b** Endothermic; [1 mark] **c** $-14\,°C$ [1 mark]

3. a For [1 mark] each: 2, (aq) and (g).

b The concentration of the acid [1 mark] and the volume of the acid [1 mark]. Also accept the actual apparatus.

c Going down the column: 0.8, 1.6, 1.5, 3.2, 4.0; [2 marks] for all correct but [1 mark] if 1 wrong.

d As the amount of magnesium metal used increases, the bigger the temperature change. [1 mark]

Reaction profiles

1. a Correctly completed profile [1 mark]; correct arrow for energy change [1 mark]; activation energy label. [1 mark]

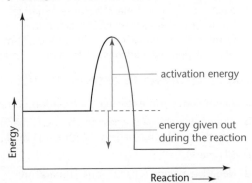

b When two particles collide many do not have sufficient energy [1 mark]; they do not meet the activation energy requirement. [1 mark]

c There would be an explosion. [1 mark] The energy from the naked flame would be transferred to the reacting particles of oxygen and hydrogen. [1 mark] So when the particles collide the activation requirement is met and the reaction takes place. [1 mark]

2.

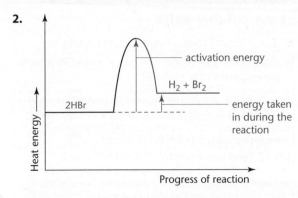

Energy change of reactions

2.

Marks awarded for this type of answer are determined by the Quality of Written Communication (QWC) as well as the standard of your scientific response.
0 marks: No relevant content.
Level 1 (1–2 marks): There is a brief description of exothermic and endothermic reactions.
Level 2: Several reasons for and against are listed.
Level 3 (5–6 marks): There is a clear, balanced and detailed description of exothermic and endothermic reactions.

Indicative content

- Bond breaking is an endothermic process.
- The activation energy is the minimum amount of energy required to break bonds.
- Bond making is an exothermic process.
- The overall energy change is a combination of both processes.
- For an endothermic reaction, the energy needed to break the existing bonds in the reactants is higher than the energy released in forming new bonds in the products.
- During an endothermic reaction, the temperature decreases as heat energy is taken in from the surroundings.
- For an exothermic reaction, the energy released from forming new bonds in the products is higher than the energy needed to break the existing bonds in the reactants.
- During an exothermic reaction, the temperature increases as heat energy is released into the surroundings.

Cells and batteries

1. a Any soluble salt, e.g. potassium nitrate. [1 mark]

b Because of the relative reactivities of the two metals. [1 mark] Zinc is more reactive than copper [1 mark] so it forces the copper ions to accept electrons. [1 mark]

c Zinc; [1 mark] $Zn + Cu^{2+} \rightarrow Cu + Zn^{2+}$ [1 mark]

d It decreases. [1 mark]

e It increases. [1 mark]

f They have the same reactivity so there is no potential difference. [1 mark]

2. Any 4 of the following points for [1 mark] each, but both types of batteries must be compared.

Non-rechargeable
Cheap to buy.
High cost/performance ratio, i.e. expensive to use in the long run.
Can only be used once.
Disposal creates lots of chemical waste.
Output falls gradually with time.

Rechargeable
Expensive to buy.
Low cost/performance ratio, i.e. cheap to use in the long term.
Can be used many times.
Disposal of fewer batteries means less waste.
Output stays constant until almost flat.

Fuel cells

1. a For [1 mark] each, any four of:

When hydrogen is burned the product is water – no pollutants are produced [1 mark], whereas when fossil fuels burn, the greenhouse gas carbon dioxide is produced. [1 mark] Incomplete combustion produces toxic carbon monoxide and particulates of carbon which can damage buildings and cause health problems. [1 mark]

Fossil fuels are non-renewable, but there is a constant supply of hydrogen. [1 mark]

Hydrogen fuel cells are efficient because chemical energy is tranformed directly into electrical energy. [1 mark] Whereas fossil fuels are less efficient because energy is transferred to the surroundings. [1 mark]

b Negative electrode: 4 and 4; [1 mark] Positive electrode: 4 and 2 [1 mark]

c The fuel is oxidised in the cell to create the potential difference that 'pushes' the electrons round the external circuit. [1 mark] At the negative electrode the released electrons move through the wire [1 mark] to the positive electrode, while the positive ions travel through the electrolyte where they are reduced by reacting with oxygen. [1 mark]

Section 6: Rates of reaction

Measuring rates of reaction

1. a A rubber bung has been put in the conical flask. [1 mark] The mass of the flask would remain constant because no gas would escape. [1 mark]

b 80.570 at 45 s. [1 mark] Simply recorded incorrectly because nothing else was changed. [1 mark]

c $80.906 - 80.686 = 0.220$ g [1 mark]

d $12 + (2 \times 16) = 44$ g. [1 mark] Moles produced = $0.220 \div 44 = 5 \times 10^{-3}$ [1 mark]

Calculating rates of reaction

1. a Rate $= 60 \div 7$ [1 mark] $= 8.57$ [1 mark] cm^3/min. [1 mark]

b

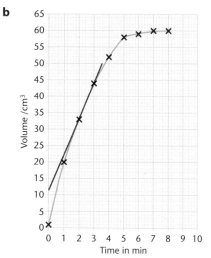

Correctly labelled axes. [1 mark] Correct scale. [1 mark] Points plotted correctly. [2 marks] Smooth curve drawn through points. [1 mark]

c

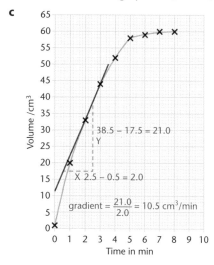

[4 marks]

Effect of concentration and pressure

1. a It will increase [1 mark]. Increasing the pressure increases the chances of successful collisions. [1 mark]

b Time = number of collisions ÷ frequency [1 mark] $= (1 \times 10^5) \div (5 \times 10^{10}) = 2 \times 10^{-6}$ s [1 mark]

c Number of collisions = collision frequency × time $= (5 \times 10^{10}) \times 2.5$ [1 mark] $= 1.25 \times 10^{11}$ [1 mark]

2. a More concentrated: the graph goes up more steeply at the start. [1 mark] And finishes earlier at the same volume. [1 mark]
More dilute: The graph goes up more gently at the start [1 mark]. The final volume of gas is less than original. [1 mark]

b Chemical reactions happen when particles collide with sufficient energy. [1 mark] Increasing the concentration of the acid increases the number of reacting particles, which increases the chances of successful reactions taking place. [1 mark] Therefore the reaction takes places at a different rate. [1 mark]. For example, if the acid is in excess, and the amount of magnesium is the same, the same volume of gas is produced as before [1 mark]. The reaction stops when all the magnesium particles are used up or all the acid has reacted. [1 mark]

Effect of surface area

1. **a** D [1 mark]

   b Rate A = 0.80 cm³/s. [1 mark] Rate C = 0.2 cm³/s. [1 mark] The rate of A was 4 times the rate of C. [1 mark]

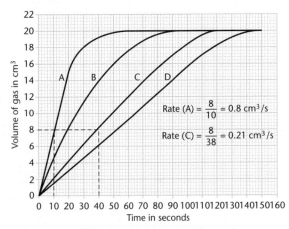

Rate (A) = $\frac{8}{10}$ = 0.8 cm³/s

Rate (C) = $\frac{8}{38}$ = 0.21 cm³/s

   c Chemical reactions take place when particles collide with sufficient energy. [1 mark] The particles in reaction A are smaller than in reaction C and therefore have a larger surface area. [1 mark] Increasing the surface area of a reactant increases the number of reacting particles, which increases the chances of successful reactions. [1 mark] Therefore the reaction goes at a quicker rate [1 mark].

   d Volume of gas produced = 20 cm³ [1 mark] = 20 ÷ 1000 = 0.02 dm³. [1 mark] Number of moles = (1 ÷ 24) × 0.02 = 8.33 × 10⁻⁴. [1 mark]

Effect of temperature

1.

Marks awarded for this answer will be determined by the Quality of Written Communication (QWC) as well as the standard of your scientific response.
0 marks: No relevant content.
Level 1 (1–2 marks): There is a brief description of the practical procedure.
Level 2 (3–4 marks): There is some description of the practical procedure.
Level 3 (5–6 marks): There is a detailed description of the practical procedure.

Indicative content

- Required apparatus: conical flask, stopclock, paper with black cross, two 25 cm³ measuring cylinders, thermometer, water baths set at 20, 30, 40, 50 and 60 °C.
- Chemicals: dilute hydrochloric acid, sodium thiosulfate.
- Control variables: concentration of acid, concentration of sodium thiosulfate.
- Method: measure out 20 cm³ of dilute acid, pour it into a conical flask; measure out 20 cm³ of sodium thiosulfate, pour it into the same flask, stand the flask on the cross and time until the cross disappears. Repeat to check the data.
- Repeat the method using reagents that have been previously warmed in the water bath to the correct temperatures.
- Record all the data in a table.

2. **a** Experiment 4: 11.5 and 0.087. [1 mark] Experiment 5: 10.0 and 0.100. [1 mark]

   b Correct scales. [1 mark] Correct labels. [1 mark] Correct points plotted. [2 marks] Line of best fit drawn. [1 mark]

   c Increasing the temperature increases the rate. [1 mark]

Effect of a catalyst

1. Rate; [1 mark] not used up during the reaction. [1 mark]

2. **a**

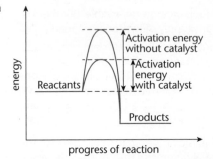

   b They create a new pathway for the reaction [1 mark] which has a lower activation energy. [1 mark]

3. Group A measured the activation energy for the uncatalysed reaction. [1 mark] Group B measured the activation energy for the catalysed reaction, in which it is lower. [1 mark]

4. To speed up the rate of reaction. [1 mark] This means they can operate at lower temperatures and pressures, [1 mark] which in turn makes the process cheaper and greener because it uses less energy in the heating stages, [1 mark] creating less greenhouse gas pollution [1 mark].

Reversible reactions and energy change

1. 2 [1 mark]

2. **a** $CuSO_4 \cdot 5H_2O(s) \rightarrow CuSO_4(s) + 5H_2O(l)$
   Correct formulae. [1 mark] Correct balancing. [1 mark] Correct state symbols. [1 mark] (accept (g) for water.)

b For the reaction to work it must be heated, [1 mark] which means that energy is being transferred into the system. [1 mark]

c Water must be added to the white copper sulfate [1 mark]

d White crystals turn blue. [1 mark] Steam is given off [1 mark]

e -78.22 kJ/mol. [1 mark] The same amount of energy is transferred out in the reverse reaction as was transferred in during the forward reaction. [1 mark]

Equilibrium and Le Chatelier's principle

1. Closed system. [1 mark] Forward and reverse reactions occur at exactly the same rate. [1 mark]

2. To the left. [1 mark]

3. **a** To the right. [1 mark]

   b It is lost [1 mark] because carbon dioxide gas escapes from the system. [1 mark]

4. The effect of changing conditions on a system that is in equilibrium. [1 mark]

5. $NaCl(s) \rightarrow NaCl(aq)$ **or** $NaCl(s) \rightarrow Na^+(aq) + Cl^-(aq)$ [1 mark]

   Solid sodium chloride is dissolving into aqueous sodium ions and chloride ions [1 mark] at the same rate [1 mark] as the aqueous sodium ions and chloride ions are combining to form solid sodium chloride. [1 mark]

Changing the position of equilibrium

1. **a** 2; [1 mark] 2 [1 mark]

   b 200 atm [1 mark]. For the forward reaction to occur, 3 volumes of gas are reduced to 2 volumes of gas. [1 mark] Applying Le Chatelier's principle, increasing the pressure favours the forward reaction. [1 mark]

   c There is a compromise to be found between the position of equilibrium, the rate of reaction, the yield and the cost. [1 mark] At low temperatures the rate of formation of sulfur trioxide is too slow to be viable. [1 mark] It is very expensive to operate at high pressures. [1 mark] The position of equilibrium at the chosen temperature will be near enough right to give an acceptable yield. [1 mark]

2. **a** Increasing the temperature, increases the yield of the forward reaction. [1 mark] Therefore the forward reaction is endothermic [1 mark] because putting more energy into the system shifts the position of equilibrium towards the right [1 mark].

   b As the pressure increases, the % conversion to **Y** decreases. [1 mark] For the forward reaction, 1 volume of gas increases to 2 volumes so the position of equilibrium favours low pressure. [1 mark] Therefore lowering the pressure will favour the forward reaction. [1 mark]

3.

Marks awarded for this answer will be determined by the Quality of Written Communication (QWC) as well as the standard of your scientific response.
0 marks: No relevant content.
Level 1 (1–2 marks): There is a brief description of equilibrium.
Level 2 (3–4 marks): There is some description of equilibrium.
Level 3 (5–6 marks): There is a detailed description of equilibrium.

Indicative content

- The Haber process reaction is $N_2(g) + 3H_2(g) \rightarrow 2NH_3(g)$
- The forward reaction is $N_2(g) + 3H_2(g) \rightarrow NH_3(g)$
- The reverse reaction is $NH_3(g) \rightarrow N_2(g) + 3H_2(g)$
- At equilibrium the rate of the forward reaction equals that of the reverse reaction. This means that ammonia is being made and decomposed at the same rate.
- At equilibrium the concentrations of the reactants and products do not change.
- Changing the position of equilibrium changes the relative amounts of reactants and products in the mixture at any one time.
- If the equilibrium lies to the left the concentrations of $N_2(g)$ and $3H_2(g)$ are higher than that of $NH_3(g)$.
- If the equilibrium lies to the right the concentrations of $N_2(g)$ and $3H_2(g)$ are lower than that of $NH_3(g)$.
- The position of equilibrium can be changed by changing the conditions of pressure and temperature.
- A catalyst does not change the position of equilibrium – only the rate of reaction.
- Increasing the pressure shifts the equilibrium to the right and more $NH_3(g)$ is produced because 4 volumes of gas are converted to 2 volumes of gas.
- The reaction is exothermic, so decreasing the temperature shifts the equilibrium to the right because less energy needs to be put into the reaction.

Section 7: Organic chemistry

Crude oil and hydrocarbons

1. **a** Hydrogen; [1 mark] Carbon [1 mark]

   b A natural fuel formed millions of years ago [1 mark] from the remains of plankton and other biomass compressed in mud. [1 mark]

   c The Sun: plants take in energy from the Sun during photosynthesis. [1 mark] Fossil fuels come from the decay of plants or the animals that feed on them. [1 mark]

   d When the supplies run out they cannot be replaced. [1 mark]

Answers

Structure and formulae of alkanes

1. **a** C_4H_{10} [1 mark]
   b Each carbon atom forms four single bonds. [1 mark] The bonds are C–H or C–C. [1 mark]

2. **a** C_2H_6 [1 mark]
   b Each time a carbon atom is added into an alkane two more hydrogen atoms are also added. [1 mark] The molecule with 4 carbon atoms has 10 hydrogen atoms, so the one with 5 carbon atoms has 12 hydrogen atoms. [1 mark]
   c C_7H_{16} [1 mark]

3. Molecule D is the odd one out [1 mark]. It has six carbon atoms and all the others have five. [1 mark]

Fractional distillation and petrochemicals

1. **a** Crude oil is heated so that all the hydrocarbons are in the gaseous state. [1 mark]
   The vapours are fed into a fractionating tower, which is hot at the bottom and cooler at the top. [1 mark] Near the bottom, hydrocarbons with higher boiling points condense and the liquids are piped off. [1 mark] Hydrocarbons with lower boiling points are still gases, so they rise up the column. [1 mark] As they move up, the hydrocarbons condense at the point where the temperature of the column is just below their boiling point. [1 mark]
   b Bitumen has much larger molecules than kerosene; [1 mark] so there are larger forces of attraction between the bitumen molecules than between the smaller kerosene molecules. [1 mark] This means that bitumen molecules are harder to separate, and so more energy is required to pull bitumen molecules away from each other; [1 mark] this is achieved at the higher boiling point.

2. Well **A** produces $(15 \div 100) \times 150 = 22.5$ kg; [1 mark] **B** produces $(20 \div 100) \times 142 = 28.4$ kg [1 mark]. So well **B** produces the most paraffin. [1 mark]

Properties of hydrocarbons

1. **a** Boiling point 320–450, Viscosity 9–10, Flammability 1–2. Accept numbers in these correct ranges for [1 mark] each.
   b Flammability decreases as the number of carbon atoms in the molecules increases. [1 mark]
   c The larger the molecules, the higher the boiling point. [1 mark]
   d The larger the molecule, the more viscous the fraction becomes [1 mark]. This is because the forces of attraction between the molecules increase as the size of the molecule increases. [1 mark]. So it becomes harder for the molecules to slide past each other. [1 mark]

Combustion of fuels

1. **a** $C_6H_{14} + 9.5O_2 \rightarrow 6CO_2 + 7H_2O$ [1 mark]

   b $C_3H_8 + 5O_2 \rightarrow 3CO_2 + 4H_2O$ Correct formulae, [1 mark] correct balancing. [1 mark]
   c When the air hole is open, there is a good supply of oxygen, so complete combustion of methane gas occurs. [1 mark] Methane burns with a clean flame to produce carbon dioxide and water. [1 mark] When the air hole is closed, the supply of oxygen is limited, so incomplete combustion of methane gas occurs. [1 mark] The products include some soot, which also burns making the flame look yellow. [1 mark]

2. Marks should be awarded for discussion around each point listed, to a maximum of 4.
   A fuel suitable for cooking when camping needs to:
   - give out enough heat energy to cook the food. [1 mark]
   - be easy to ignite at lowish temperatures. Some fuels need to be primed or have a fuse, which can be difficult to use. [1 mark]
   - be readily available and affordable so that people can easily purchase it. [1 mark]
   - burn with a clean flame so that the pans do not get covered with soot and it does not produce toxic gases such as carbon monoxide. [1 mark]
   - have a low carbon footprint, to reduce greenhouse gas emissions which contribute towards global warming. [1 mark]

Cracking and the alkenes

1. **a** **A** is $C_{18}H_{38}$; [1 mark] **B** is aluminium oxide or porous pot [1 mark]
   b When large alkane molecules are converted into smaller, more useful alkene molecules. [1 mark]
   c 2; [1 mark] C_4H_8, C_3H_6 and C_2H_4. [1 mark]
   d Add a few drops of bromine water to the gaseous product. [1 mark] If the bromine water decolourises then alkenes are present; [1 mark] and the reaction has been successful. If it remains orange-brown then only alkanes are present. [1 mark]

2. Hot vapours of large hydrocarbon molecules are mixed with steam to a very high temperature. [1 mark] Thermal decomposition occurs, breaking the large molecules into smaller molecules. [1 mark] It is an important process because the demand for smaller hydrocarbons, such as petrol, is higher than for those containing larger hydrocarbons. [1 mark] Smaller alkanes are used as fuels, while the more reactive alkenes are important starting materials for making polymers and other petrochemicals. [1 mark] Accept answers illustrated with an equation.

Structure and formulae of alkenes

1. **a** C_3H_6 [1 mark]
   b Every alkene molecule contains a C=C double bond [1 mark] The naming prefix 'meth' means there is only 1 C atom in the molecule. [1 mark] A molecule with just 1 carbon atom cannot have a C=C double bond, therefore it doesn't exist. [1 mark]

2. **a** C_4H_8 [1 mark]

   b Each time a C is inserted in an alkene, 2 more hydrogen atoms are also added. [1 mark]

   The molecule with 5 carbon atoms has 10 hydrogen atoms, so the one with 6 carbon atoms has 12 hydrogen atoms. [1 mark]

   c $C_{10}H_{20}$ [1 mark]

3. Unsaturated molecules are reactive because they contain a C=C double bond. [1 mark] Saturated molecules contain only C–C single bonds, making them much less reactive. [1 mark]

Reactions of alkenes

1. **a** Compound D [1 mark]

   b Compound A [1 mark]; Compound C [1 mark]

2. **a** 2C; [1 mark] 2 [1 mark]

   b It is from the carbon or soot. [1 mark]

3.

Marks awarded for this answer will be determined by the Quality of Written Communication (QWC) as well as the standard of your scientific response.
0 marks: No relevant content.
Level 1 (1–2 marks): There is a brief comparison of structure and chemical reactions.
Level 2 (3–4 marks): There is some comparison of structure and chemical reactions.
Level 3 (5–6 marks): There is a detailed comparison of structure and chemical reactions.

Indicative content

- The formula of ethene is C_2H_4; the general formula of alkenes is C_nH_{2n}.
- The formula of ethane is C_2H_6; the general formula of alkanes is C_nH_{2n+2}.
- The functional group of alkenes is C=C. They are called unsaturated hydrocarbons. The double bond makes them reactive.
- The functional group of alkanes contains C–C single bonds. They are called saturated hydrocarbons because they are unreactive.
- Both alkenes and alkanes burn in oxygen to produce carbon dioxide and water. In both cases incomplete combustion occurs if the oxygen supply is limited.
- Alkenes react with hydrogen, water and the halogens by the addition of atoms across the C=C bond so that it becomes a C–C bond. You should include displayed formulae showing reactants and products.
- Alkanes don't react with hydrogen or water. They are not very reactive with halogens without the presence of strong UV light (post-16 idea).

Structure and formulae of alcohols

1. Compound A [1 mark]

2. **a** ethanol + carbon dioxide [1 mark]

   b $C_2H_4(g) + H_2O(g) \rightarrow C_2H_5OH(l)$. [1 mark] for correct formulae; [1 mark] for correct state symbols.

   c The reaction is catalysed by enzymes found in yeast. [1 mark] If the temperature is too high the enzymes are denatured and if too cold they will be inactive and the reaction rate will be slow; [1 mark] the temperature needs to be in the range 25–40 °C for the yeast to work efficiently. [1 mark] Similarly using pH values outside the optimum range will denature the active site of the enzyme. [1 mark]

Uses of alcohols

1. **a** Food flavouring; [1 mark] Adhesives [1 mark]

   b $C_2H_5OH + 3O_2 \rightarrow 2CO_2 + 3H_2O$. [1 mark] for correct formulae; [1 mark] for correct balancing.

   c Mix an alcohol and a carboxylic acid [1 mark] in the presence of a concentrated acid catalyst. [1 mark] The reaction is alcohol + carboxylic acid $\rightarrow$ ester + water. [1 mark] For example, ethyl ethanoate is formed from ethanol and ethanoic acid. [1 mark]

2.

Marks awarded for this answer will be determined by the Quality of Written Communication (QWC) as well as the standard of your scientific response.
0 marks: No relevant content.
Level 1 (1–2 marks): There is a brief description of how the problem was solved.
Level 2 (3–4 marks): There is some description of how the problem was solved.
Level 3 (5–6 marks): There is a clear description of how the problem was solved.

Indicative content

- **A** is an alkane, $C_{10}H_{22}$, which underwent cracking.
- **B** is an alkene because it is a 'reactive product' of cracking.
- **B** has 2 carbon atoms so it must be C_2H_4.
- **C** is the other product of cracking so must be an alkane.
- **C** is C_8H_{18} because this uses up the remaining carbon atoms and hydrogen atoms from the cracking reaction.
- The cracking reaction is $C_{10}H_{22} \rightarrow C_8H_{18} + C_2H_4$.
- **D** must be an alcohol because alcohols react with sodium to produce hydrogen gas, which pops when lit.
- **D** is C_2H_5OH because it is made from **B**, which has 2 carbon atoms.
- **D** is formed by the reaction $C_2H_4 + H_2O \rightarrow C_2H_5OH$.
- So the equation for the reaction that produces hydrogen is $2C_2H_5OH + Na \rightarrow 2C_2H_5ONa + H_2$.

Carboxylic acids

1. **a** Compound C [1 mark]

   b Butanol [1 mark]

2. **a** It fizzes. [1 mark]

   b H^+ ions make solutions acidic. [1 mark] In strong acids all the molecules undergo complete ionisation when dissolved in water, but in weak acids the ionisation is not complete. [1 mark] Nitric acid is strong because only H^+ and NO_3^- ions exist in solution. [1 mark] Methanoic acid is weak because not all the methanoic acid molecules dissociate to form H^+ and HCO_2^- ions. [1 mark]

3.

Marks awarded for this answer will be determined by the Quality of Written Communication (QWC) as well as the standard of your scientific response.
0 marks: No relevant content.
Level 1 (1–2 marks): There is a limited comparison.
Level 2 (3–4 marks): There is some comparison with examples.
Level 3 (5–6 marks): There is a detailed comparison with clear examples.

Indicative content

- The formula of ethanoic acid is CH_3COOH.
- The formula of ethane is C_2H_6; the general formula is C_nH_{2n+2}.
- The functional group of carboxylic acids is –COOH.
- The functional group of alkanes contains only C–C single bonds.
- All alkanes are insoluble in water, whereas some of the smaller carboxylic acids are soluble in water and the rest are partially soluble.
- Both carboxylic acids and alkanes burn in oxygen to produce carbon dioxide and water. In both cases, incomplete combustion occurs if the oxygen supply is limited.
- Alkanes are very unreactive with other chemicals.
- Carboxylic acids are weak acids with pH values around 4–6; alkanes are neutral.
- Carboxylic acids are more reactive – they react with sodium metal to produce a salt and hydrogen gas; they also react with metal carbonates and metal hydroxides. This is due to the presence of the –COOH group.
- Carboxylic acids will react with alcohols to produce an ester and water, but there is no reaction with alkanes.

Addition polymerisation

1. **a** Ethene monomers are used to make poly(ethene). [1 mark]

    b

    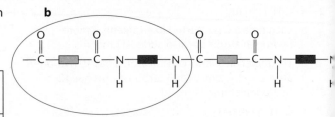

    [1 mark] for each.

    c The carbon–carbon double bond in the propene molecules opens up [1 mark] and is replaced by a single carbon–carbon bond as the monomers join together [1 mark] during the addition polymerisation reaction.

Condensation polymerisation

1. **a** Monomers with two different functional groups join together to form a polymer. [1 mark] During the reaction a small molecule (e.g. water) is eliminated. [1 mark]

    b

    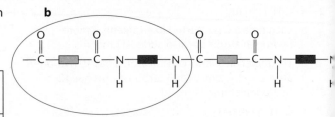

    c [1 mark] for the link; [1 mark] for the unit; [1 mark] for n

2. Poly(ethene) is made by addition polymerisation whereas polyester is made by condensation polymerisation. [1 mark]. Each process involves joining small monomer molecules to make a large polymer molecule. [1 mark] In the case of polyethene there is one ethene monomer in which the double bond opens up during the reaction, whereas polyester is made using two monomers [1 mark] with different functional groups. During a condensation reaction a water molecule is eliminated from the –OH and HOOC– groups. [1 mark]

Amino acids

1. **a** Circle the $-NH_2$ group [1 mark] and the –COOH group. [1 mark]

    b Water [1 mark]

    c $(-HNC_2H_4CO-)_n$ [1 mark]

2. Amino acids react by condensation polymerisation to produce polypeptides. [1 mark] For example, glycine polymerises to produce the polypeptide poly(glycine). [1 mark]

    $nH_2NCH_2COOH \rightarrow (-HNCH_2COO-)_n + nH_2O$ [1 mark]

    A polypeptide can be made from different amino acids. [1 mark] Proteins are made when different polypeptides join up. [1 mark]

DNA and other naturally occurring polymers

1. Glucose. [1 mark]

2. A condensation polymerisation reaction [1 mark] takes place in which water molecules are eliminated. [1 mark] One end of the glucose molecule loses –OH [1 mark] while the other end loses –H. [1 mark]

3. **a** It carries all the genetic instructions needed for living things to grow and function. [1 mark]

    b DNA molecules have two polymer chains made from four different monomers called nucleotides. [1 mark] The polymer chains are joined together in the shape of a double helix. [1 mark]

    c From top to bottom: phosphate, sugar, base. [3 marks]

    d Circle the nucleotide (to include the sugar and phosphate group) with 'T'. [1 mark]

Section 8: Chemical analysis

Pure substances, mixtures and formulations

1. **a** Pure distilled water boils at 100 °C [1 mark] but if the water is impure the boiling point will be slightly higher. [1 mark]

    b It has been contaminated. [1 mark]

    c The impure water is evaporated to dryness. [1 mark]

    d It is impure [1 mark] because it has a broad melting point range. [1 mark]

    e Carry out a series of chemical and/or instrumental tests. [1 mark]

2. **a** So they will work. [1 mark] Too much active ingredient could make a patient worse; not enough might not make them better. [1 mark]

    b To make sure the formulation is correct. [1 mark]

    c Sample **D** [1 mark] because it contains approximately 5% more paracetamol than all the other samples. [1 mark]

Chromatography

1. **a** 4 [1 mark]

    b **S** next to top spot. [1 mark]

    c Amino acids are more soluble in ethanol than in water. [1 mark]

2. **a** He will not get good results because he has made several mistakes.

    • The solvent level is above the start line [1 mark]. He needs to put it below to avoid the ink spots smudging at the start. [1 mark]

    • He has used pen ink to label the spots. He must use pencil because this will not smudge. [1 mark]

    • He has spilt some ink in the middle of the chromatography paper. This will start to separate when the solvent reaches it. [1 mark]

    b It is made up of four different dyes. [1 mark]

    c Calculate the R_f values [1 mark] and compare them to data in reference tables [1 mark] (or repeat using standard dyes and see if the spots match).

Chromatography and R_f values

1. **a** The ratio [1 mark] of the distance moved by the spot to the distance moved by the solvent. [1 mark]

    b $R_f = \dfrac{\text{distance moved by spot}}{\text{distance moved by solvent}}$

2. **a** $\text{distance moved by solvent} = \dfrac{\text{distance moved by spot}}{R_f}$

    $= \dfrac{18}{0.36}$ [1 mark] $= 50\,\text{mm}$ [1 mark]

    b C = 5.5 mm; [1 mark] D = 28 mm [1 mark]

3. **a** The top horizontal line. [1 mark] A line drawn from the baseline to the centre of the green spot. [1 mark]

    b $R_f = 4 \div 6$ [1 mark] $= 2.5$ [1 mark]

    c It is a ratio of two lengths, so the units cancel each other out. [1 mark]

Tests for common gases

1. **a** Turns damp blue litmus paper red. [1 mark]

    b Chlorine [1 mark]

2. **a** Gas **Z** is carbon dioxide [1 mark] because it turns limewater cloudy. The other test results confirm this because when dissolved in water CO_2 is acidic, which is shown by a positive litmus test [1 mark]. The other two tests confirm that neither oxygen nor hydrogen is present because they give negative results. So, solid **Y** must be a metal carbonate because an acid + metal carbonate → a salt + water + carbon dioxide. [1 mark] Further tests will be needed to identify the metal. [1 mark]

    b

    Correct set up; [2 marks] Correct labels. [1 mark]

Flame tests

1. **a** Potassium chloride [1 mark]

    b Lithium [1 mark]

2. She is using the yellow Bunsen flame instead of the blue one; [1 mark] so is just seeing the Bunsen flame colour. [1 mark] Or the nichrome wire is dirty. [1 mark]. It has lots of sodium ions stuck to it so the other flame colours don't show up. [1 mark]

3. **a** Copper carbonate [1 mark]

    b 2; [1 mark] State symbols: s, aq, aq, l and g in that order. [2 marks] [1 mark] if one is wrong.

Metal hydroxides

1. **a** For [1 mark] each: **A** is Na^+; **B** is Ca^{2+}; **C** is Cu^{2+}.

    b Calcium carbonate [1 mark]

    c Further tests are needed to identify these anions [1 mark] because they both gave a negative test for carbonate ions. [1 mark]

2. The ion will be calcium or magnesium. [1 mark] Carry out a flame test: if the flame is red, the ion is calcium. [1 mark]

3. **a** A pale blue precipitate forms. [1 mark]

    b $CuSO_4 + 2NaOH \rightarrow Cu(OH)_2 + Na_2SO_4$ Correct reactants. [1 mark] Correct products. [1 mark] Correctly balanced. [1 mark]

4. a Fe^{2+} [1 mark]

    b   2; [1 mark] $Fe(OH)_2$. [1 mark]

Tests for anions

1. Add a few drops of acidified silver chloride to some of each solution. [1 mark] If a white precipitate appears then chloride ions are present. [1 mark]

    Add a few drops of dilute hydrochloric acid and then barium chloride to some of each solution. [1 mark]. If a white precipitate appears then sulfate ions are present. [1 mark]

2.

Marks awarded for this answer will be determined by the Quality of Written Communication (QWC) as well as the standard of your scientific response.
0 marks: No relevant content.
Level 1 (1–2 marks): There is a brief description of some of the required reactions and expected observations.
Level 2 (3–4 marks): There is brief description of the required reactions and expected observations.
Level 3 (5–6 marks): There is a clear and detailed description of the required reactions and expected observations.

Indicative content

- All three samples can be identified using just two tests and a third to confirm the result.
- Carry out a flame test on each sample.
- A yellow flame is observed for sodium ions and an orange-red flame for calcium ions.
- Carry out a test for chloride ions by adding acidified silver nitrate to a solution of each sample.
- A white precipitate will be observed if chloride ions are present.
- To check your answer, carry out a test for sulfate ions by adding dilute hydrochloric acid and then barium chloride solution to the sample you think is sodium sulfate.
- A white precipitate will be observed if you are correct.

Instrumental methods

1. a   **F** [1 mark]

    b   **B**; [1 mark] and **C** [1 mark]

    c   It is more accurate; [1 mark] and more sensitive. [1 mark] The results are obtained more quickly. [1 mark]

2. It is a quick way of identifying people who were present at the scene of a crime. [1 mark] The test results are compared to those held in a database to see if there is a match. [1 mark]

3. a   Ethene; [1 mark] and propene. [1 mark]

    b   Add bromine water. It will change from yellow to colourless if an alkene is present. [1 mark]

    c   The chemical test simply indicates the presence of an alkene. [1 mark] But the mass spectrum can identify the actual alkenes present. [1 mark] The mass spectrum method is also more accurate and you can also work out the relative amounts of each molecule present. [1 mark]

Flame emission spectroscopy

1. a   The sample is put into a flame [1 mark] and the light given out is passed through a spectroscope. [1 mark]

2. Any one from: You can measure the concentration of the ions. You can identify different ions in a mixture. The results are more accurate. [1 mark]

3. a   Rubidium. [1 mark] The spectral lines are at the same wavelengths. [1 mark]

    b   Run a sample of solution **Y** through the flame emission spectrometer; [1 mark] analyse the result. If there is only one spectral line observed at 580 nm then only sodium ions are present. [1 mark] If several lines are observed then it has been contaminated. [1 mark]

    c   The flame colours for sodium (yellow) and potassium (lilac) are quite close together. [1 mark] It will be hard to tell if any potassium is present because the colour will be masked by the intense yellow of the sodium. [1 mark]

Section 9: Chemistry of the atmosphere

The Earth's atmosphere – now and in the past

1. a   Oxygen [1 mark]

    b   Carbon dioxide or water vapour or any noble gas. [1 mark]

2. a CO_2 is 95/100 = 19/20; All other gases are 5/100 = 1/20; [1 mark] The ratio is 19 : 1. [1 mark]

    b   78.1% of the Earth atmosphere is nitrogen = 78/100. [1 mark] 3% of the Mars atmosphere is nitrogen = 3/100 [1 mark] 78 ÷ 3 = 26 times more nitrogen on Earth. [1 mark]

    c   96.5 ÷ 0.039 = 2474.36 [1 mark] = 2470 [1 mark]

3.

Marks awarded for this answer will be determined by the Quality of Written Communication (QWC) as well as the standard of your scientific response.
0 marks: No relevant content.
Level 1 (1–2 marks): There is a brief comparison of the evidence.
Level 2 (3–4 marks): There is some comparison of the evidence.
Level 3 (5–6 marks): There is a detailed comparison of the evidence.

Indicative content

- Scientists were not present millions of years ago to make direct measurements, so it is difficult to be certain that the theory is correct.

- It is difficult to make accurate measurements from the evidence available, which means there is more uncertainty in the model.
- The theory is based on the composition of gases vented out of volcanoes today, making the assumption that volcanoes today give out the same composition of gases as volcanoes did millions of years ago.
- At the start of this period, Earth's atmosphere may have been like that of Mars or Venus today – i.e. lots of carbon dioxide and not much oxygen. Scientists make models and theories based on this data collected today from different planets.
- As the oceans formed, carbon dioxide dissolved in the water and carbonates were precipitated. Over millions of years the carbonates were converted into sedimentary rocks.
- Scientists now collect further evidence by measuring carbon and boron isotope ratios in sediments under the sea. This type of evidence is valuable because it can be linked directly back through time.
- Supporting evidence found in rocks and from cores of ice formed millions of years ago add more weight to the theory.
- Proxy evidence, such as counting the number of stomata found in fossilised leaves, to make assumptions about the levels of carbon dioxide in the atmosphere is also used to help develop scientific models.
- Data becomes more reliable as different teams of scientists around the world find that they have independently produced similar data.

Changes in oxygen and carbon dioxide

1. **a** Human life didn't exist billions of years ago, so there was no one to make direct measurements. [1 mark]

   b Evidence from fossil records. [1 mark] Evidence in sedimentary rock formations. [1 mark]

   c 6, 6, 6; [2 marks] if all three are correct; [1 mark] if one is wrong.

   d About 2.5 billion years ago. [1 mark]

   e Initially the oxygen levels started to rise but then remained at about 3% for the next billion years; [1 mark] before they started to rise steadily over the next billion years to reach approximately 20% today. [1 mark] However, some scientists think that the oxygen levels could have peaked at about 33% about 0.27 billion years ago, before dropping back to the 20% seen today. [1 mark]

2. | Marks awarded for this answer will be determined by the Quality of Written Communication (QWC) as well as the standard of your scientific response. |
   | --- |
   | 0 marks: No relevant content. |
   | Level 1 (1–2 marks): There is a brief comparison of the atmosphere past and present. |
   | Level 2 (3–4 marks): There is some comparison of the atmosphere past and present. |
   | Level 3 (5–6 marks): There is a detailed comparison of the atmosphere past and present. |

   Indicative content

   - Today's atmosphere is approximately 80% nitrogen, 20% oxygen, with small amounts of other gases such as water vapour and carbon dioxide.
   - The early atmosphere was mostly carbon dioxide, with no oxygen.
   - Early on, intense volcanic activity released gases including lots of nitrogen, water vapour with some methane and ammonia.
   - Green plants evolved and produced oxygen.
   - The first oxygen was used to oxidise metals such as iron.
   - Carbon dioxide was used by plants in photosynthesis.
   - The water vapour cooled and condensed forming the oceans.
   - Carbon dioxide dissolved in the water and carbonates were precipitated.
   - Over millions of years, carbonates were converted into sedimentary rocks.
   - Scientists were not present millions of years ago to make direct measurements.
   - Evidence has been found in rocks, sediment cores and ice cores formed millions of years ago.
   - It is difficult to make accurate measurements from the evidence available.

Greenhouse gases

1. **a** Nitrogen [1 mark]

   b Short wavelength radiation from the Sun enters the atmosphere. [1 mark] Some of it transfers into thermal energy when it reaches the Earth, some of it goes back into space. [1 mark] Long wavelength radiation radiated back from Earth is absorbed by greenhouse gases. [1 mark] A gradual build-up of greenhouse gases means that less heat is reflected back into the atmosphere causing an increase in temperature. [1 mark]

2. **a** Every 100 000 years it has cycled between about 280 ppm and 180 ppm. [1 mark]

   b It stayed at about 280 ppm until about 1800; [1 mark] when it started to increase rapidly until it reached about 380 ppm in 2000. [1 mark]

   c Increase = 380 − 280 = 100 ppm. [1 mark]
   % increase = (100 ÷ 280) × 100 [1 mark] = 35.7%. [1 mark]

   d Any suitable answer, e.g. The Industrial Revolution. [1 mark] Human activities like burning fossil fuels or deforestation. [1 mark]

   e An increase in global temperatures. [1 mark]

Global climate change

1. a An increase in average global temperatures. [1 mark]

   b Sea levels rise. [1 mark]

2. a So that the results can be checked and compared with those obtained in other areas of the world. [1 mark] So questions regarding, for example, the quality of the data can be answered. [1 mark]

   b It is very complex with lots of different variables. [1 mark]

3. Any two of the following: [1 mark] for an environmental change; [1 mark] for a possible impact.

- Temperature stress for humans and wildlife – too hot to make a living; will eventually die out.
- Water stress for humans and wildlife – no fresh water supplies in some areas which will lead to mass migration/species becoming extinct.
- Changes to food production capacity – different crops growing in different parts of the world; some parts may not be sustainable for farming any more.
- Changes to the distribution of wildlife – migration patterns change, which could affect local food chains and webs.

Carbon footprint and its reduction

1. The total amount of carbon dioxide and other greenhouse gases emitted over the full life cycle of a product, service or event. [1 mark]

2 For [1 mark] each: any four points to a maximum of 4 – include linked explanations.

- Double/triple-glazed windows.
- Loft/cavity wall insulation.
- Switch appliances off when not using them.
- Install solar panels.
- Use low-energy light bulbs.
- Set the thermostat a bit lower.
- Recycle products when finished with them.

3. Use carbon capture and storage systems because these take CO_2 emissions from power stations and deposit them in underground storage containers. [1 mark]
Make legislation so new buildings have to install solar panels. [1 mark]
Do a public-awareness campaign to encourage individuals to reduce their carbon footprint. [1 mark]
Accept other reasonable answers to a maximum of 3.

4. a Between 1984 and 2004 the world's population grew by about 1.5 billion people. [1 mark] The concentration of methane in the atmosphere increased from about 1625 ppb to 1750 ppb over the same period. [1 mark]. So we can conclude that as the population grew, the concentration of methane gas increased. [1 mark]

   b Grazing animals give off methane. As the population grows so does the demand for food [1 mark], so more cattle are reared to supply dairy products and meat. [1 mark]

Air pollution from burning fuels

1. For [1 mark] each, any three of: sulfur dioxide; carbon monoxide; oxides of nitrogen; solid particulates.

2. a $N_2(g) + O_2(g) \rightarrow 2NO(g)$: Correct formulae. [1 mark] Correct balancing. [1 mark] Correct state symbols. [1 mark]

   b By fitting catalytic converters. [1 mark]

3. a $S + O_2 \rightarrow SO_2$: Correct reactants. [1 mark] Correct product. [1 mark]

   b Damages buildings/statues/gravestones because it will react with the building materials. [1 mark] Damages/kills vegetation. [1 mark]

4. 2, 3, 2, 4: [2 marks] if all correct. [1 mark] if only one is wrong.

Section 10: Using resources

What does the Earth provide for us?

1. a Finite sources from the pie chart are: gas, coal, nuclear and 'other'. [1 mark] = 30 + 30 + 19 + 1.8 = 80.8% [1 mark]

   b It will decrease. [1 mark] These finite sources will be used up and be replaced by more alternative renewable energy sources. [1 mark]

2. a Accept either answer, so long as the arguments given support it.

   Cotton because it comes from a natural source; [1 mark] and uses less energy during production; [1 mark] than polyester. Or: Polyester because although it comes from a finite source, during production it uses a lot less water; [1 mark] and produces less CO_2 than the production of cotton; [1 mark] which also relies on the use of pesticides.

   b Similar data associated with use, e.g. washing; [1 mark] and data associated with the end of its useful life. [1 mark]

3. a $CaCO_3(s) \rightarrow CaO(s) + CO_2(g)$

   b Limestone is a finite resource. [1 mark] Quarrying leaves an environmental scar and causes noise pollution. [1 mark] Cement production uses a lot of energy because high temperatures are required. [1 mark] CO_2, a greenhouse gas, is also released as a by-product. [1 mark] Although cement comes from a natural resource it is not very sustainable. [1 mark

Answers

Safe drinking water

1. **a** Accept a poorly drawn diagram so long as it clearly shows the sample being heated. [1 mark] A condenser. [1 mark] An open system and where the pure water is collected. [1 mark]

   b Evaporation [1 mark] followed by condensation. [1 mark]

   c First filter it [1 mark] to remove the solid particles making it cloudy. [1 mark] Then distil it [1 mark] to remove any dissolved salts present. [1 mark]

   d For [1 mark] each, any two of: Check to see if its boiling point is 100 °C, which is the boiling point of pure water. Check to see if its pH is 7. For sample **A** he could carry out a chloride ion test to see if any chloride ions were present.

2. Pure water contains only H_2O molecules [1 mark]. Potable water is safe to drink [1 mark]. Potable water may contain dissolved salts. [1 mark]

3. It is done to kill any microbes present in the water. [1 mark] If carried out at an earlier stage in the process, microbes could be reintroduced in a later stage. [1 mark]

Waste water treatment

1. **a** So that each household can treat their own waste water. [1 mark]. It is too expensive to link them into the sewage network used in urban areas. [1 mark]

   b So it can be emptied when it is full of sludge. [1 mark]

   c It leaks out of the tank onto the land [1 mark] where it rejoins the natural water cycle. [1 mark]

2. **a** For [1 mark] each, two of: remove solid waste, organic matter, harmful microorganisms and chemicals.

   b Screening removes large solids and grit. [1 mark] Sedimentation produces sewage sludge and effluent. [1 mark] Biological treatment of effluent to make the water safe. [1 mark] Sterilisation makes sure there are no dangerous microorganisms present. [1 mark]

   c Biological treatment tank: aerobic digestion takes place – reactions involve oxygen. [1 mark] Sludge tank: anaerobic digestion takes place – no oxygen is required for the reactions. [1 mark]

   d Fertilisers; [1 mark] as a renewable fuel. [1 mark]

Alternative methods of extracting metals

1. **a** Phytomining uses plants to absorb metal compounds which are then harvested. [1 mark] The plants are burned to produce an ash that contains metal compounds which are then extracted using electrolysis or displacement reactions. [1 mark]

   b The copper ions that have been absorbed by the plant will react with oxygen in the air as the plant burns. [1 mark] $Cu^{2+} + O^{2-} \rightarrow CuO$. [1 mark]

   c $CuSO_4(aq)$ [1 mark] + $H_2O(l)$ [1 mark]

   d Add iron to the copper sulfate solution. [1 mark] A displacement reaction will take place leaving the copper. [1 mark] Filter the solution – copper will be left in the filter paper. [1 mark]

2.
Marks awarded for this answer will be determined by the Quality of Written Communication (QWC) as well as the standard of your scientific response.
0 marks: No relevant content.
Level 1 (1–2 marks): There is a brief description of some of the issues surrounding the extraction of copper.
Level 2 (3–4 marks): There is some description of the issues surrounding the extraction of copper.
Level 3 (5–6 marks): There is a clear description of the issues surrounding the extraction of copper.

   Indicative content
   - Most of the high-grade copper has already been mined, leaving low grade copper ore.
   - The Earth's resources are limited, so we need to extract all we can.
   - Traditional methods of mining low-grade ore are not economically viable and have a huge environmental impact – such as noise, dust, destroys natural habitat.
   - Bioleaching is the extraction of specific metals from their ores using bacteria.
   - Bioleaching is done by producing leachate solutions containing metal compounds.
   - For example, copper compounds are then extracted by displacement by scrap iron or by electrolysis.
   - Bioleaching is simple, cheap and can extract from low-grade ores.
   - Bioleaching is environmentally friendly and produces less waste gas.
   - One disadvantage of bioleaching is that it is slow.
   - However, toxic chemicals are produced and it is not very efficient because there are lots of stages.

Life cycle assessment

1. **a** To assess the environmental impact of the products. [1 mark]

   b It looks at the materials and the energy involved in making the product. [1 mark] How it is used. [1 mark] What happens to the product at the end of its useful life. [1 mark]

2. Yes, because it has only examined one small part of the life cycle of the materials. [1 mark] The raw materials for window frames come from crude oil, which is a finite resource, whereas wood is renewable [1 mark]. Plastics have many manufacturing stages that require high temperatures, whereas once the trees have grown it is easy to get the wood. [1 mark] At the end of their life, aluminium frames can be recycled, whereas plastic usually goes to landfill. [1 mark] Accept any other relevant points up to 4 in total.

Answers

3.

Marks awarded for this answer will be determined by the Quality of Written Communication (QWC) as well as the standard of your scientific response.
0 marks: No relevant content.
Level 1 (1–2 marks): Little of the LCA data has been used and applied.
Level 2 (3–4 marks): Some of the LCA data has been used and applied.
Level 3 (5–6 marks): The LCA data has been fully used and applied to the problem.

Indicative content

- The raw materials for plastic cups are finite and will eventually run out.
- Raw materials needed for paper cups are renewable, provided that the trees are replaced at the same rate as they are cut down.
- There are several stages to the production of plastic cups that must be carried out at high temperature and pressure. This will have high energy costs and high emissions of greenhouse gases if fossil fuels are the power supply.
- The production of paper has much lower energy costs because it is carried out at atmospheric pressure and low temperatures.
- The use of chemicals during the production of paper cups can be harmful to the environment if released into water supplies or on land after use.
- Plastic cups can be reused more often than paper cups, which should mean fewer are needed.
- Plastic cups that go to landfill will remain there because they are not biodegradable, whereas paper cups are.
- Paper cups are more sustainable than plastic cups because of differences in the raw materials and manufacturing processes.

Ways of reducing the use of resources

1 Aluminium is a reactive metal [1 mark] that is extracted by electrolysis, [1 mark] which needs a lot of energy.

2. a Overall in the UK it increased. [1 mark]

b [1 mark] for any three points: Overall usage in the UK, England and Wales increased. Usage in Scotland and Northern Ireland decreased. A 5p charge per bag was introduced in Scotland in 2014, so fewer bags were bought. But the population of Scotland is small compared to that of England, so an overall impact was not really seen.

c $7.64 \times 0.15 = 1.15$ billion. [1 mark]

d There is a saving in the amount of raw materials used to make the bags – crude oil. [1 mark] There are savings in the energy required for processing, which means lower greenhouse gas emissions. [1 mark] Fewer bags are thrown away in bins and also in hedgerows, waterways etc. which can harm wildlife. [1 mark] Accept other reasonable answers.

Corrosion and its prevention

1. a Cover it in oil [1 mark] to keep out oxygen and water. [1 mark]

b If the pier steel supports are scratched or damaged the magnesium reacts and provides protection [1 mark] for the iron because it is more reactive. [1 mark]

c They wear away and need replacing. [1 mark]

d 2 [1 mark]; 3 [1 mark]

e Zinc is less reactive. [1 mark]

2. Set up the following test tubes:

A with 2 cm^3 water, an iron nail and leave open. [1 mark] **B** with 2 cm^3 water, an iron nail and anhydrous calcium chloride bung in the top (to dry the air by removing any water vapour). [1 mark] **C** with 10 cm^3 water, an iron nail, a layer of oil to stop air getting in. [1 mark]

Note the observations at the start and after a week. [1 mark]

Alloys as useful materials

1. a 10% [1 mark]

b 8% [1 mark]

c In pure copper, the copper layers slip over each other easily. [1 mark] As aluminium atoms are introduced they disrupt the pattern, making it harder for the layers to move resulting in an increase in strength. [1 mark] After the percentage of Al reaches a maximum, it becomes easier for the layers to move over each other because some of the aluminium atoms may start to align themselves. [1 mark]

2. a $(88 \div 100) \times 1.3 = 1.1440$. [1 mark] So the spearhead contains 1.14 kg of copper. [1 mark]

b In pure copper, the atoms are arranged in layers [1 mark]. It is soft because the layers can slide over each other [1 mark]. In bronze, the tin atoms disrupt the structure [1 mark] stopping the layers from sliding over each other. [1 mark] So bronze is harder than pure copper. Accept diagrams showing the structure of copper and bronze.

Ceramics, polymers and composites

1. a Composite [1 mark]

b Concrete is weak in tensile strength because the bonds between the binder (cement) and the reinforcement (gravel) are not very strong. [1 mark] Adding steel rods that are high in tensile strength will increase the overall tensile strength of the concrete because the rods will take the force. [1 mark]

2. a Melamine resin as is rigid whereas the others are flexible [1 mark] and it won't melt when hot food comes straight out of the oven. [1 mark]

b In melamine there are strong cross-links between the polymer chains. [1 mark] In HD poly(ethene) there are just weak forces between the polymer chains. [1 mark] Accept a diagram.

3. Thermosoftening plastics consist of individual tangled polymer chains. [1 mark] They melt when they are heated [1 mark]. The liquid polymer can then be poured into a mould [1 mark] where it cools and solidifies to keep its new shape. [1 mark]

The Haber process

1. **a** $N_2 + 3H_2 \rightleftharpoons 2NH_3$: Correct formulae of reactants. [1 mark] Correct formula of products. [1 mark] Balanced. [1 mark]

   b The reaction is reversible. [1 mark]

   c Increases as the pressure increases. [1 mark]

   d The reaction involves 4 moles/volumes of gas changing to 2 moles/volumes of gas. [1 mark] Increasing the pressure favours the forward reaction because it reduces the volume. [1 mark]

   e The percentage decreases as the temperature increases. [1 mark]

   f The reaction is exothermic. [1 mark] A lower temperature favours the forward reaction. [1 mark]

   g Low temperatures give higher yields but the reaction rate is very slow. [1 mark] Working at very high pressures gives a high yield but the equipment is very expensive. [1 mark] The chosen conditions give a yield of about 15% ammonia, but the reactants are recycled through a continuous flow system. [1 mark] Need to use a compromise between position of equilibrium, rate of reaction, yield and cost. [1 mark]

Production and use of NPK fertilisers

2.

Marks awarded for this answer will be determined by the Quality of Written Communication (QWC) as well as the standard of your scientific response.
0 marks: No relevant content.
Level 1 (1–2 marks): There is a brief description of some of the issues surrounding the sustainability and environmental impact of the processes.
Level 2 (3–4 marks): There is some description of the issues surrounding the sustainability and environmental impact of the processes.
Level 3 (5–6 marks): There is a clear description of the issues surrounding the sustainability and environmental impact of the processes.

Indicative content

- Most fertilisers are ammonium- and/or nitrate-based.
- Ammonia made by the Haber process requires high temperatures and pressures – lots of energy is required for this.
- The Haber process may have a really high carbon footprint if non-renewable energy sources such as fossil fuels are used – the impact of this is increased greenhouse gas emissions that lead to global warming. Using renewable energy sources such as wind or solar will be more sustainable in future.
- Nitric acid is manufactured from ammonia.
- The raw materials for the Haber process – nitrogen from distillation of air and hydrogen from natural gas or water – means that we need to consider the environmental impact of extraction processes.
- Phosphate rock is mined and used to make fertilisers. Mining leaves scars on the landscape. Mining areas are often dusty, with noise pollution and increased road traffic. They also cause damage to the environment.
- Raw materials will eventually run out.
- For increased sustainability we should consider where factories are sited – close to feedstock production to reduce costs, source of power etc.